# Cartography

## an introduction

### SECOND EDITION

## Giles Darkes and Mary Spence MBE

### British Cartographic Society

First published in the UK by the British Cartographic Society in 2008
c/o Royal Geographical Society
1 Kensington Gore
London SW7 2AR

Written by Giles Darkes and Mary Spence MBE
Designed by Mary Spence MBE

The authors and the British Cartographic Society would like to thank all Corporate Members, cartographers and others who have contributed maps and illustrations to this publication.



ISBN 978-0-904482-24-9



**British Cartographic Society**
Registered Charity Number 240034
**www.cartography.org.uk**

# Contents

<table>
<tr><td>1</td><td>2</td></tr>
<tr><td>3</td><td>4</td></tr>
</table>

1  **London Tourist Map** © Cosmographics

2  **Lake District Walking Map** © Harvey Maps

3  **United Kingdom Wall Map** © Tiger Moon

4  **London DataShine Web Map** © Oliver O'Brien

# Foreword

THE BRITISH CARTOGRAPHIC SOCIETY — a group of people united by an enthusiasm for all things to do with maps — has long aimed to share the expertise of its members. In particular, we strive to share knowledge of how to produce good maps with a wider mapmaking community.

It is often asked why we need to produce new maps. 'Surely everything has been mapped now?' is a question often heard. In fact, more people are making and sharing maps than ever before, and we encounter maps perhaps more often than we might realise. With access to online as well as printed maps and the unprecedented availability of data in all forms, more people make or see maps than at any point in the past, and the trend is set to continue.

However, in today's data rich but information-poor societies, communicating the right message is all the more important. Cartography is the art of geovisualisation, and requires creative solutions. Having an understanding of what maps are and how they are made can only serve to improve the quality of map design, and to engender an appreciation for the fact that a map is more than the sum of its parts.

Although there are a number of excellent cartography textbooks available, none of them is aimed at the curious beginner, or the GIS practitioner tasked with producing good maps but with no formal training in how to do so. We hope this book fills that gap with its direct and pragmatic approach, looking at a wide range of aspects of maps and mapping, and not specific to any software package. This new, expanded edition, written by two leading practitioners with decades of experience, will provide you with a wealth of practical information and help you to make better maps.

Above all, I hope that you will enjoy putting the wisdom of its pages to good use.

Dr Alexander Kent
President of the British Cartographic Society
Canterbury, UK
Summer 2017

# Introduction

CARTOGRAPHY IS ALL ABOUT MAPS. It covers many aspects of maps, from their history and collection to their design and production.

This book has been written to introduce the subject, and in particular to explain some of the basic principles that underlie the creation of a good map. Many people make maps but the number of mapmakers who have formal training in cartography is few. By looking at different aspects of what makes for a successful map and by offering practical suggestions, the aim is to help mapmakers to understand the principles of good map design.

The book is also designed to introduce other aspects of cartography including their history and classification and to explore the basics of data quality, data preparation, and the presentation of thematic data on maps.

In whatever way you're involved with maps, we hope that this brief guide will help you to understand them better.

<table>
<tr><td>1</td><td>2</td><td>3</td></tr>
<tr><td>4</td><td>5</td><td>6</td></tr>
<tr><td>7</td><td>8</td><td>9</td></tr>
</table>

1  St Aubyn Estates: Porthgwarra © Clear Mapping Company

2  Illustrated Inverness Street Map © LovellJohns

3  Antique Style World © Global Mapping

4  Durham Interactive © GD3D & CC3D

5  Geological Map of the World © The Future Mapping Company

6  Discover Cairngorms National Park © Stirling Surveys

7  Colour-blind Friendly Political Map of the World © Cosmographics

8  Historical Map of York about 1850 © The Historic Towns Trust

9  Bedmap 2: Bed Topography of the Antarctic © British Antarctic Survey

A MAP IS A GRAPHIC SUMMARY of the wider world. It is a distillation of geography where elements of the world around us are presented as symbols, scaled down and simplified to make them understandable. The relative position and the nature of features in the wider world is summarised in order to communicate a message.

Maps vary greatly in their graphic forms, from rigorous, scientific documents to the purely imaginative images of the artist. However, most maps attempt to convey a factual message, portraying aspects of real-world geography and the practical parts of this book are aimed at those making conventional maps of the real world.

There are four graphic elements which appear on most maps: points, lines, polygons (or areas), and text in the form of labels. Most maps have all these elements, but even if one of the first three is missing, the presence of labels is essential. When looking at an aerial or satellite image such as seen on Google Earth™, common sense and a basic knowledge of geography allows recognition of the nature of the features that you can see: a lake, building or road, for example. However, in order to add meaning and context to the image, each feature needs a name: Lough Neagh, Holyrood Palace, Route 66. If the image is of a familiar area, you are able unconsciously to add names to the features but if the area is unfamiliar, labels are essential.

Images with added names are a half-way step to a true map. The final transition is made when the features seen in the image are portrayed solely as symbols. The process of mapmaking involves the classification of real-world objects (for example lines may be classified as roads or railways), the selection of what to include on the map and what to leave out (generalisation), and the choice of graphic object to depict the real-world feature at a smaller scale (symbolisation). The final stage is the addition of labels to add specific meaning to the symbols.

**Different types of maps:**

LEFT:
London: street map.
© CollinsBartholomew

CENTRE:
World Petroleum
Congress: thematic map.
© LovellJohns

RIGHT:
The Road to Milford
Sound (3-D): tourist map.
© Geographx NZ

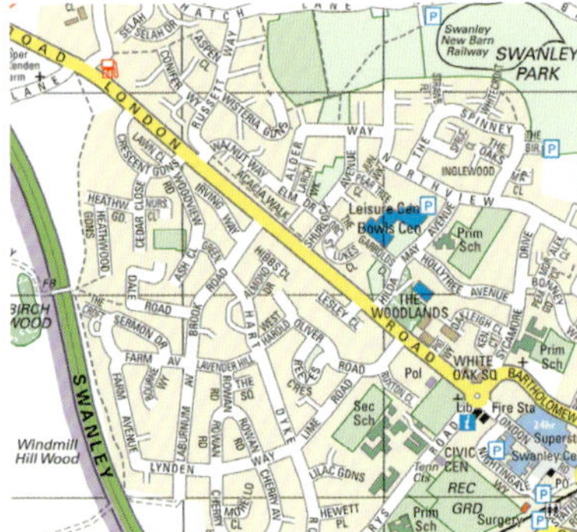

CARTOGRAPHY HAS BEEN DESCRIBED as the art, science and technology of mapmaking. Although relatively few people earn a living by producing maps, almost everyone is involved with maps as users. Growing numbers of people make their own maps using websites, the Global Positional System (GPS) or a geographical information system (GIS). For many, maps are regarded as special and something to be cherished — works of art valued beyond the sum of their contents.

Cartography is the study and practice of the many facets of maps and mapmaking. The core of cartography is the process of making maps, but as well as their design and practical production, cartography encompasses a range of activities associated with maps and mapping: studying their history and meaning; their classification; their printing, distribution and sale; and their collection, conservation, and curation in map libraries.

Within the wider field of map production, there are many specialisms. Cartography includes the design and production of marine, air and military charts, statistical maps, planning maps, walking and orienteering maps, geological maps, tourist and travel maps, weather and climate maps, general and specialist atlases, cartograms, and transport network diagrams – to name but a few. If you create maps, work with them or use them, you're closely linked to cartography.

### did you know...?

The term "Cartography" was coined in 1859, from the Latin, *carta* meaning papyrus or paper and *-graphy*, from the Greek, meaning to write, or to draw.

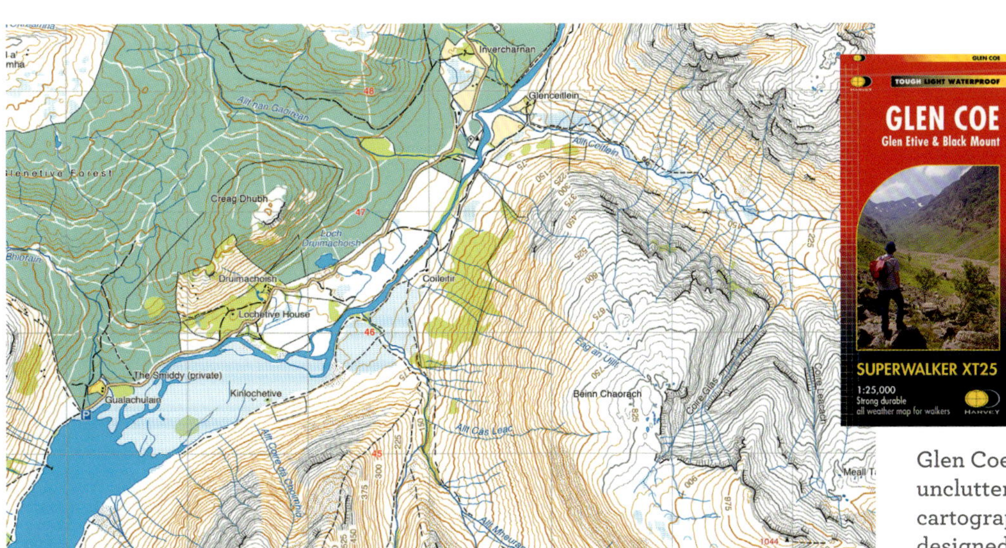

Glen Coe: Clear, uncluttered, traditional cartography, specifically designed for walkers.

© Harvey Maps

MAPS HAVE BEEN AROUND for 8,000 years or so — certainly longer than the written word. The date of the earliest maps is debatable because it is hard to define exactly what a map is, and whether early art really was intended to be a map. It is perhaps surprising to note that for many centuries mapping was more concerned with measuring and depicting the shape of the globe, and of large regions of it, than of local areas.

Portolan chart by Jorge de Aguiar, 1492: The oldest known signed and dated chart of Portuguese origin.

Courtesy: Beinecke Rare Book and Manuscript Library, Yale University, New Haven, USA

### Data gathering

Take the methods of gathering mappable data, for example. The earliest geographers used sailors' and travellers' reports of the locations and distances to places to compile lists such as Ptolemy's *Geography*. These weren't turned into maps until centuries later. Eventually, surveyors produced maps from their own observations and travels. But it was the development of mathematics and then scientific instruments — the compass, the plane table and optical devices such as the alidade, telescope, staff, sextant and ultimately the theodolite and level — which have led to the systematic measurement of the landscape. Since the early 20th century, aerial photography and now satellite imagery have allowed the rapid survey of large areas of land.

Most maps are compiled from a wide variety of different sources which are brought together to create the final picture. Historical maps may be scanned and georeferenced to a national grid, for example, so that other data can be fitted on top, thus affording a direct comparison between past and present. Old, inaccurate mapping can be improved and updated with new data within a GIS, often creating an interactive product.

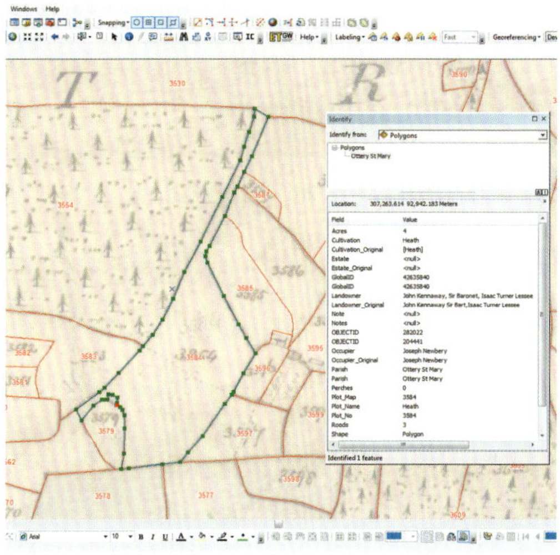

GIS data capture from historic mapping.
Courtesy: LovellJohns

## Map drawing

The first maps were drawn by hand using brushes, ink and parchment and were limited in number, unique and of high value. It was the invention of the printing press that allowed the distribution of many copies of one map. Wood engravings gave way to copper printing plates. Lithographic and photomechanical means of making printing plates were developed, meaning that more maps could be produced consistently and cheaply. Printed colour was added, while paper improvements meant that fine lines and lettering could be printed.

Earth Platinum: The largest atlas in history, published in 2012. Leather-bound, weighs 150 kilograms, 128 pages each measuring 1.8m high x 1.4m wide. Only 31 copies exist. The aim in producing the atlas was to create a benchmark in cartography and to leave a legacy for future generations.
Courtesy: Millennium House, Australia

UAV hexacopter — ideal for capturing small-format aerial imagery.

Courtesy: Aibotix

## The technical revolution

The greatest revolution has been since the 1970s as computer technology has joined up the strands of data gathering, map creation, reproduction, distribution and map use in a way unimaginable in the past. Mapping has developed links with image processing, visualisation and spatial analysis as well as with graphic production and publishing. Surveyors now routinely use digital methods to gather and record data about geographical features. Maps are compiled, designed, printed, viewed and used digitally and may never be printed. GIS has a digital map at its core, and web-mapping, SatNav, and mobile phone apps are examples of the widening ways maps are used in a digital environment.

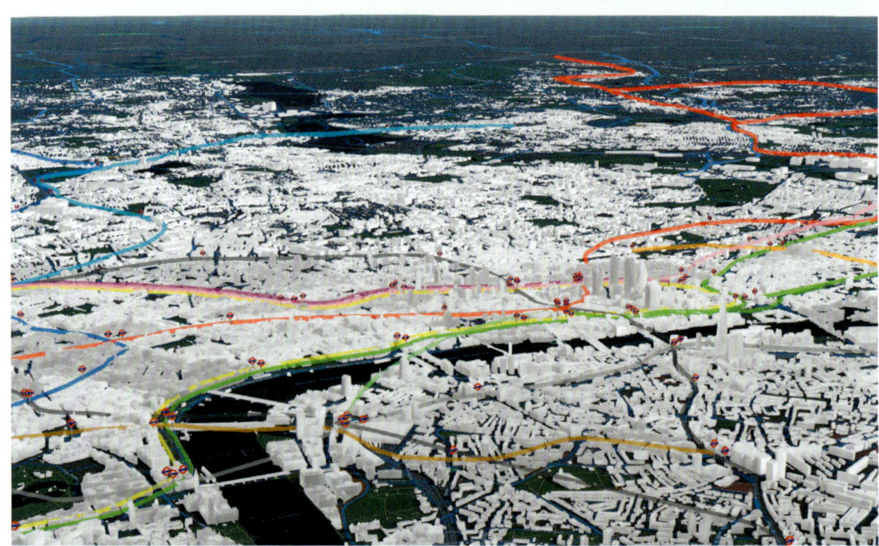

3-D visualisation: GD3D proBuildings overlaid on a custom basemap, London underground created as 3-D geometries.

Contains Ordnance Survey data © Crown copyright and database right 2017, © GD3D & CC3D 2017

## Map design

Changing designs of maps and map symbols reflect the growth of graphic design in general as well as the continuing search by cartographers to find better ways to show ever-more varied geographical data. General-purpose reference maps have been joined by specific-purpose thematic maps requiring a different approach to design. Maps prepared in colour have become the norm and a variety of typefaces have allowed for the graphic differentiation of features by label. It is now much quicker to create a map and distribute it widely than ever it was in the past. However, although the platforms for map creation and delivery have radically altered, it is worth emphasising that the principles of good map design have not fundamentally changed.

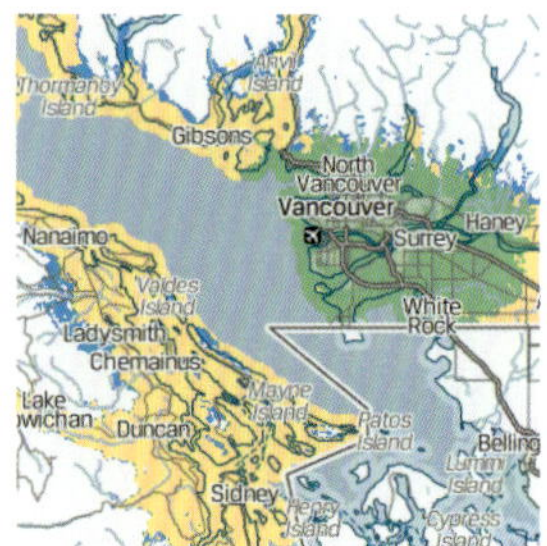

Mobile phone coverage.

© Europa Technologies

**A**LL MAPS HAVE IN COMMON the fact that they show geographical features, but they vary dramatically in design and presentation. A map in a travel brochure may apparently bear little resemblance to one showing population statistics, but both are depicting *locations* and *attributes*.

What counts as a map includes a wide range of graphics, presented on different media: paper or electronic delivery such as computer screen, television screen, mobile phone, and personal navigation equipment. Map types include physical, political, and general-purpose topographic, maritime and air charts, thematic maps, atlases, road maps, transport maps and diagrams, photomaps (with aerial photographs or satellite images in the background), 3-D models, cartograms (which deliberately distort map shapes and areas to show statistics in a different light) and pictorial maps. While static maps are unchanging, dynamic maps are designed to change through time: fly-throughs, SatNav maps, GPS receivers and many maps designed for web delivery or mobile apps are in this category, updating as the map reader moves in space or changes scale.

Although the range of maps is clearly large, a significant proportion of them can be classified as either *topographic* or *thematic*. Topographic maps are essentially reference maps showing a set range of general features in the landscape — Ordnance Survey® Landranger™ and Explorer™ maps are well-known examples. These maps aim to indicate the character of the terrain and symbolise a wide range of features that are considered useful to most users. Related to topographic maps are maritime and air charts which emphasise features of relevance to navigators against a general topographic background.

OS Maps app on desktop, tablet and smartphone.
Courtesy: Ordnance Survey

Legible London wayfinding map (left)
Courtesy: Applied Information Group

Legible London monolith (right).
Wikimedia Commons, Stevekeiretsu cc-by-sa-4.0

Thematic maps emphasise a specific topic — e.g. population, health, or planning zones — against a reduced topographic background. Thematic

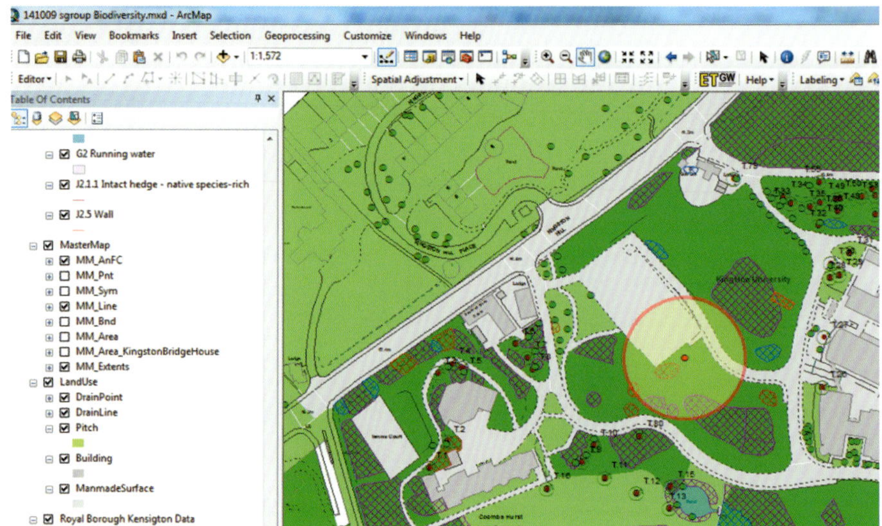

Kingston upon Thames:
Online biodiversity
mapping.

Courtesy: LovellJohns

maps are especially diverse and include tourist maps, land cover and land use maps, planning, climate, and weather maps. The many statistical maps come in this category. They also include maps for walkers, cyclists and orienteers which use general topographic features as the background, but emphasise specific features of use to their intended audience.

City of Vancouver:
Online tourism mapping,
mastered in four scales.

Design and development
by Applied Wayfinding

The classification of maps is inevitably arbitrary and it is easy to cite examples of maps that do not fall readily into being either topographic or thematic — pictorial maps, for instance, or mental maps or maps of imaginary lands. However, the subject is introduced here to emphasise the diversity of map types that exist, and their varying graphic nature.

Although it seems trite to say, maps have as many functions as they have readers. Two readers may use exactly the same map for totally different purposes. But maps are used because they convey information easily and, for many purposes, are the best way to visualise spatial information. Given that perhaps 90% of all information has a geographical component to it (*where* something exists is a feature of it), it is not surprising that mapping it may be an effective way of communicating something about it.

Some of the main categories of map function are:

- for navigation and mobility
- for the location of places
- for analytical purposes, involving measurement
- to show the geographical extent of a class of information
- to show statistics or trends
- to visualise a landscape
- to stimulate spatial understanding.

More generally, the function of a map is to inform. It is a useful shorthand way of summarising complex spatial information. However, a well-designed map also has an intrinsic appeal to our curiosity and to the imagination.

Pictorial wall map:
Attention All Shipping.
© Jane Tomlinson

## did you know...?

The London Underground diagram is a map which shows the correct relationship between stations and lines, but distorts real distances between them. Central London is greatly increased in size relative to the suburbs. It has served as the model for maps of transport networks throughout the world.

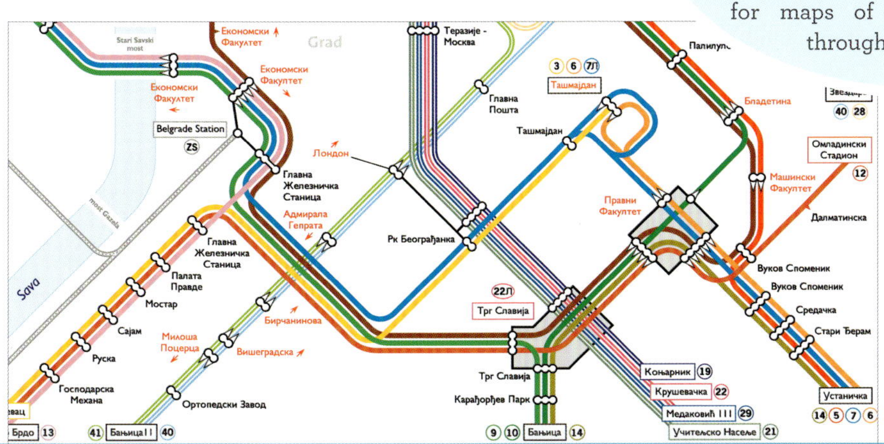

Transport mapping:
Belgrade Tram and
Trolleybus.
© Communicarta

**GPS surveyor.**

Courtesy: Ordnance Survey

The length of a base line A–B is measured very carefully on the ground. The position of point C is obtained by measuring the angles ABC and BAC and using trigonometry to work out the lengths A-C and B-C. Once C's position is known, the position of point D can be calculated.

SURVEYING IS THE TECHNIQUE of determining the positions of points on the earth's surface by measuring the distances and angles between them. Most large-scale maps and plans are made or updated by surveying.

The standard instruments used by a surveyor include an accurate measuring tape, a theodolite for measuring angles, a level for determining differences in height, a device for electronic distance measurement (EDM) and a GPS receiver. The functions of a theodolite and a level are usually now combined in a total station (TS) — an instrument which combines EDM with the ability to measure both horizontal and vertical differences in angle between two points.

The relative positions of a number of points can be determined through the process of triangulation. This involves the accurate measurement of a base line (a straight, level line between two known points). Once this line has been determined, observations are taken of a third point from each end of the base line. If the relative angles of the third point from each of the base points are noted, the position and distance of the third point from the base line can be determined through trigonometry. Once the position of the third point has been calculated, the location of a fourth and subsequent points can be worked out by taking observations of angle from any two of the known points. Vertical changes in level can be measured at the same time. In small areas, the effect of the curvature of the earth can be ignored (called plane surveying) but over larger areas or long distances, the effect is taken into account (geodetic surveying).

Triangulation can be used over great distances and most countries have a basic triangulated network covering their territories. In the UK, the network is marked with triangulation pillars, although only a few of these are now regularly maintained. Bench marks are added to maps and to features on the ground to indicate a point whose precise position and elevation has been worked out.

The advent of the satellite navigation systems such as the Global Positioning System (GPS) has revolutionised surveying. Information transmitted from satellites allows a GPS receiver to calculate its location and height above sea level. Great accuracy (a few millimetres) can be achieved with repeated measurement from GPS signals. Many TSs use GPS to work out their position in terms of a national grid or latitude and longitude, and the positions of the points surveyed from it are automatically determined.

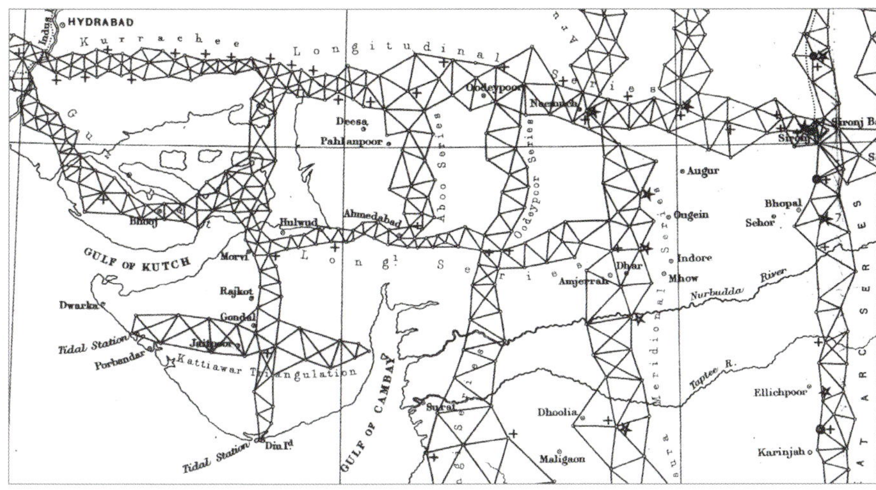

Extract from the index chart to the Great Trigonometrical Survey of India, 1870.

Total station theodolite.
Courtesy: ICSM Australia

Total stations record the positions of surveyed features and also their type (e.g. a corner of a building, the centre point of a road) by the surveyor adding a feature code to the point. Each real-world object has a feature code. The data collected can either be uploaded to a hand-held computer, or, increasingly, emailed to another computer. This information can be used to automatically generate a map. Higher-end total stations will show the map on-screen, and if an existing map is already loaded, the map will be automatically updated on site, allowing the surveyor to check it. If the map is tied in to a reference system such as the British National Grid, the exact location of the new features added can then be worked out with ease.

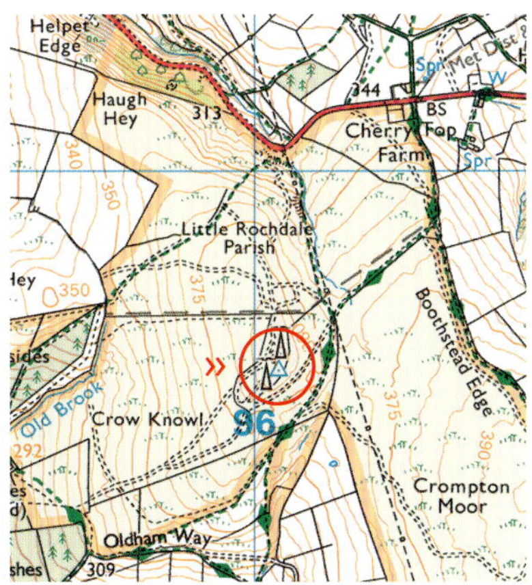

OS benchmark (above), triangulation pillar at Crow Knowl on Crompton Moor (far left)
Wikimedia Commons, J. Atherton cc-by-sa-2.5
and how it appears on the OS Explorer map.
© Ordnance Survey

LAND-SURVEYING TECHNIQUES are ideal for the detailed and accurate mapping of small areas, and are used for updating large-scale plans and cadastral maps (maps which show property boundaries). However, to survey large or remote areas takes an immense amount of time, and so photogrammetric techniques are used.

Photogrammetry is the science of taking accurate measurements from photographs. In cartography, the photographs are usually aerial, taken from a plane looking vertically down on the land. A series of overlapping photos of the ground are taken as the aircraft flies along a flight path. If the same area appears on both photographs, the image can be viewed in a *stereoscope* — a device where overlapping areas are viewed simultaneously and the brain builds a 3-D model of the landscape where the images overlap. This is known as a stereo image.

In photogrammetry, a stereo image of a portion of the earth is created within an instrument called a stereoplotter; computer software can also be used to simulate a stereoplotter. If the positions and heights of three control points on the image are already known (for example from a ground survey) then the relative position and height of any other point in the stereo image can be worked out. The image has to be carefully calibrated to do this.

Photogrammetry can be used to capture elevation data. Using a stereoplotter, a point of a particular height on the stereo image is found (e.g. a point at 100m elevation). The operator can then move a pointer around the image following all points at 100m elevation, and a contour line is recorded.

Remote sensing is a general term used to describe the observation of the earth from above. It includes sensing from aircraft, but usually it is shorthand for *satellite* remote sensing. Satellites monitoring the earth carry sensors which are capable of recording land cover either by passively detecting light or other

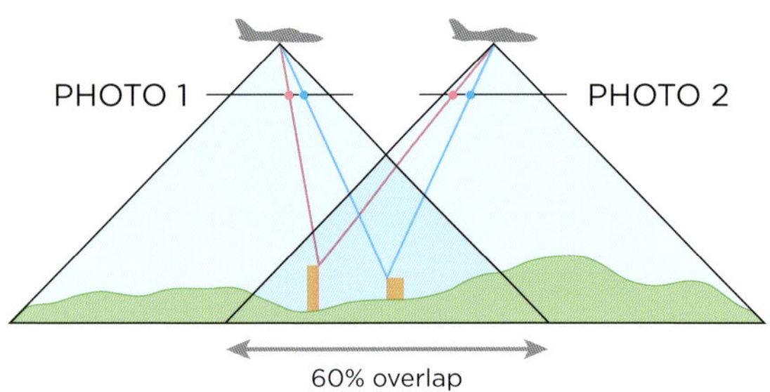


Two photographs of the same area are taken from slightly different positions, with a high degree (about 60%) of overlap between them.

OPPOSITE:
The photogrammetrist is looking at an overlapping pair of aerial photographs through stereo eyepieces. The overlap allows the land to be seen in three dimensions and detail of the terrain can be precisely plotted — vegetation, built-up areas, roads, drainage and contours.

Courtesy: Harvey Maps

radiation reflected from the earth, or by actively sending signals to the surface and measuring the reflected signal. Although many satellite images use reflected light in the visual spectrum, many sensors in other spectra (such as infrared) reveal different information, such as the health of crops or the underlying geology.

The cartographic applications of satellite imagery tend to be fewer than those of aerial photography. Stereo pairs of satellite images can be obtained and contours derived from them. The principal problem is that the resolution of the imagery (the smallest discernible object that can be seen) tends to be quite large, although resolution continues to increase and smaller objects can be seen. However, satellite images can be very useful for mapping large areas and for determining general land cover, such as vegetation.

Neither aerial photography nor satellite imagery alone can be used to make a map. Maps need labels to add meaning, and feature names don't appear in the images — a road may be visible, but the fact that it is the Stuart Highway must be found from elsewhere. Therefore 'ground data' has also to be obtained. Cloud cover limits the chances of seeing the earth and typically there are only 30 to 50 flying days a year in the British Isles when cloud-free imagery can be successfully captured.

Cadastral mapping © Danish Geodata Agency **overlaid on an orthophoto.** © Agency for Data Supply and Efficiency, Denmark and the Danish municipalities

**did you know...?**
An orthophoto is an aerial photograph that has been rectified so that the scale is uniform. Measurements can be made directly on it and it can be used in a GIS as a base to overlay other information.

Satellite imagery.
© PlanetObserver

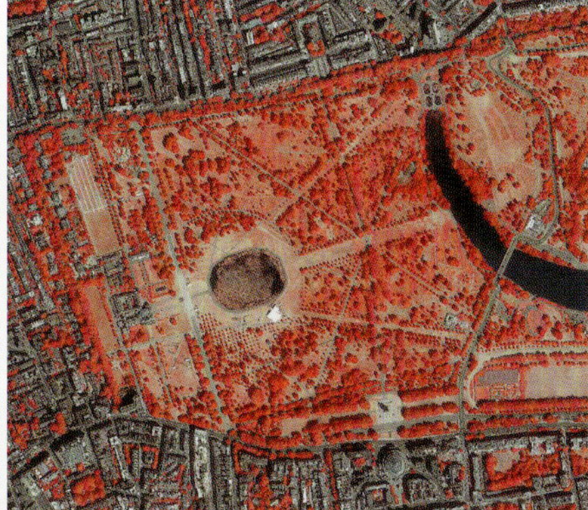

10cm resolution aerial survey over Greater London for UKMap, 2013. Full colour (far left), near infrared (NIR) (left).
© 2017 Insurance Services Offices, Inc.

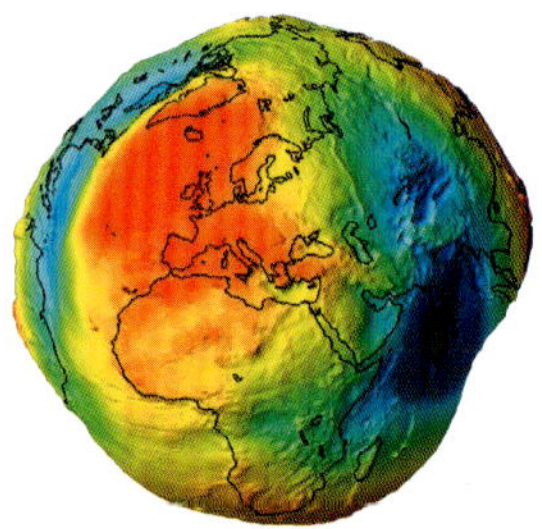

A model of the geoid, or 'earth-shaped' object. The surface depicted is the gravitational equipotential surface, or sea level. The blue colours represent areas where the geoid is below the mathematical ellipsoid and the red areas are above the ellipsoid. The differences between low and high points have been hugely exaggerated.

© European Space Agency

**M**OST MAPS HAVE a means of showing which part of the earth they are depicting. The two most common systems used on maps are latitude and longitude, and coordinates on a national grid system. Some maps use the Universal Transverse Mercator system of reference.

Latitude and longitude denote how many degrees, minutes and seconds a place is either north or south of the Equator (latitude) and east or west of an arbitrary zero line — almost always the Greenwich Prime Meridian (longitude). However, giving the lat/long of any point is not quite as straightforward as it seems because a point's position can vary according to which *horizontal datum* is being used.

The reason for this is that the earth is not a regular shape. It is approximately a sphere, but it is slightly flattened at the poles and bulges a little at the Equator. In fact, it is even more irregular in that it bulges in some places and dips in others. As a result, treating the earth as a perfect sphere and giving lat/long references as if it were is misleading. To take that into account, throughout the last 200 years different mathematical models have been devised which use an equation to simulate the shape of the earth. These are called *reference ellipsoids* or *spheroids*. Different ellipsoids take into account the flattening of the earth and provide a 'best fit' model for a particular part of the world: in Great Britain, the Airy reference ellipsoid of 1830 fits well, but poorly elsewhere on the globe. A horizontal datum defines the shape of the earth in mathematical terms, including definitions of the zero degree lines of longitude and latitude (usually Greenwich and the Equator), the diameter of the earth at the poles and the Equator, and an equation based on an ellipsoid for determining the position of any point on the earth.

With the arrival of satellite navigation systems such as the GPS, a unified global system was needed which would define the earth's shape mathematically and be a reasonable fit for all parts of the world, although a perfect fit almost nowhere. This was the World Geodetic System 1984, known as WGS84. It takes its origin as the notional centre of mass of the earth. All GPS readouts for location relate to the WGS84 datum, and increasingly maps use WGS84 for their datums, in order to relate GPS readings to map positions. Admiralty charts began changing to WGS84 in 2000.

An ellipsoid which fits well in one part of the world may not fit well in another. A local ellipsoid is defined with different size, shape and origin.

The differences in position between two points with the same latitude and longitude on difference datums can be significant. In Cornwall, a line of longitude on the system used by Ordnance Survey is about 70m west of its WGS84 equivalent, and on the eastern cost of Suffolk it rises to about 120m difference. For everyday use, this may not matter, but for a tanker navigating a narrow channel and using GPS guidance, it could be critical.

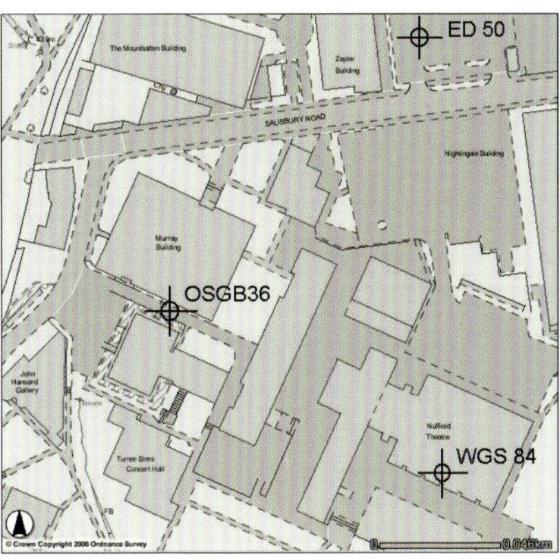


Three different points with the same latitude and longitude in three different coordinate systems — OSGB36, WGS84 and ED50.

Courtesy: Ordnance Survey © Crown copyright and database right 2017

Vertical heights on maps are usually given with reference to sea level. However, since the earth is an irregular shape which varies with different gravitational pull (basically because the density of the earth's crust varies), sea level varies across the earth. Its level depends on the gravitational pull at any point on the earth, and sea level should be thought of as a gravitational equipotential surface — a surface which has the same gravitational pull. It rises and falls considerably (by up to about 12m) from the notional surface of an ellipsoid. Sea level is therefore more complicated than it seems at first.

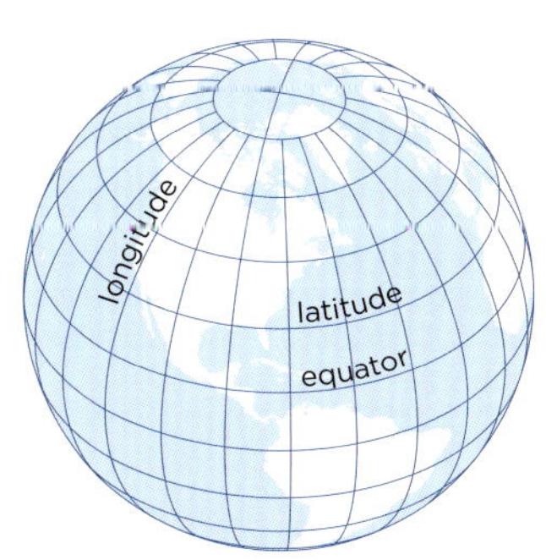


The *graticule* on a map is the depiction of the lines of latitude and longitude which invisibly surround the earth. It can either be shown as a continuous web or as a series of intersecting crosses which show the latitude and longitude values at a particular point.

A map *grid* is a reference system used to define locations in a given area. In Great Britain, the OS National Grid is a rectangular grid imposed on the curving earth, which is based on a point 49°N and 2°W. It has its origin well west of Cornwall so that all points on the grid have a positive number. Any point in GB can be given a location by stating how far east (eastings) and how far north (northings) of that point it is. Northern Ireland uses the Irish National Grid. Almost all countries in the world have their own map grids.

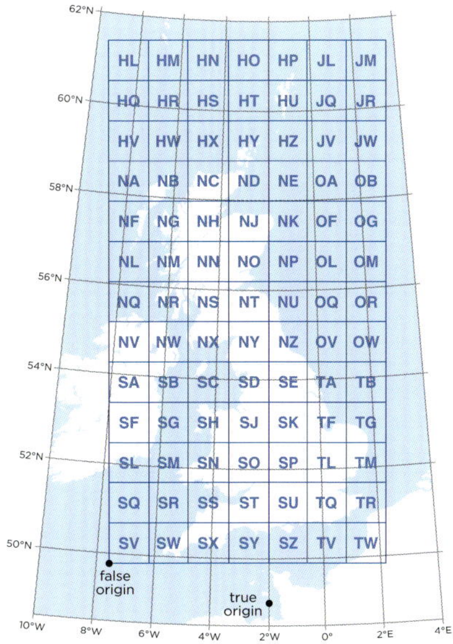


British National Grid.

Gerhard Mercator (1512-1594) was renowned for creating the 1569 world map based on a new projection which represented sailing courses of constant bearing (rhumb lines) as straight lines. The Mercator projection is widely used for maritime and air charts.

Albers Equal Area Conic is used for regions that are predominantly east-west in extent.

**T**HE EARTH IS A 3-dimensional object, but the vast majority of maps are 2-dimensional. In transferring the shape of land or oceans from the round globe to the flat map, a systematic, mathematical approach is needed to ensure that each point on the globe appears at the right point on the map. The process for doing this is known as *map projection*.

When transferring shapes from the globe onto a map, you cannot avoid introducing distortions. You can keep distances between points correct, or relative areas correct, or angles correct, but you cannot keep more than one of these properties at any one time. Map projections have different characteristics. Those that keep area correct are called *equal area* or *equivalent*; those that keep distances right are called *equidistant*; and those that keep angles correct are known as *conformal*. There is also a large group of projections which are hybrids — they combine different qualities of several projections, usually in the interests of keeping shapes familiar, but they do not keep angle, distances or areas strictly correct.

Cartographers need to use the right projection if mapping any area of the world larger than a small country. For maps of small areas, the projection chosen becomes less important because the distortions reduce as the area of the world shown gets smaller. There are no 'right' or 'wrong' projections to use — only good and poor choices for particular purposes.

For most general purposes maps of the world, you'll need to use an equal area projection. Almost all distribution maps (such as population density, land cover, or climatic regions) should use an equal area

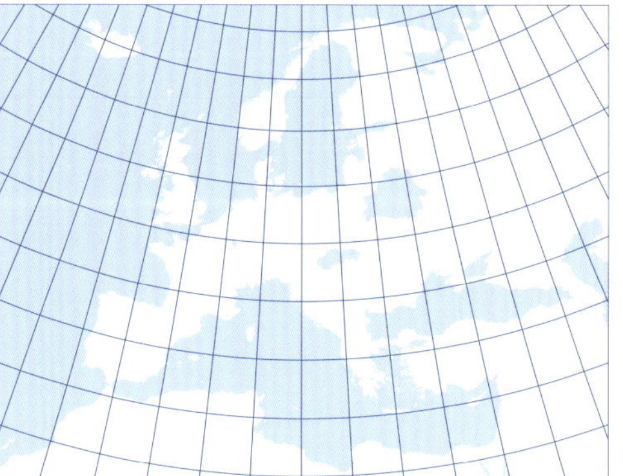

projection so that the reader sees the landmasses in their correct proportions. A non-equal area projection will unintentionally give greater importance to the data for one region just because it looks larger than it should.

For world maps, try using one of these equal area projections: *Cylindrical equal area*, *Mollweide*, *Sinusoidal*, *Eckert IV* or *Hammer-Aitoff*. Another useful projection which is not strictly equal area is the *Robinson* projection which keeps the world's shape familiar without distorting area too much. An *orthographic* projection looks like a view of the globe and can be very useful as a locator map to show where the main map is on the globe. The *Peters* or *Gall-Peters* projection, widely used by charities and campaigning organisations, although equal area, introduces shape distortions which make some regions of the world look unfamiliar. *Mercator's* projection shouldn't be used for most world maps because it distorts areas, especially in high latitudes.

If the map is to be used for navigation, then you should use a projection which keeps *angles* correct, such as *Mercator's*. If distances are to be correct then an equidistant projection is appropriate, such as a *cylindrical equidistant* or *Cassini* where distances from the equator are correct, or an *azimuthal equidistant* projection based on the North Pole which is good for maps of the northern hemisphere.

Showing the shortest distance between two places on the earth (known as a great circle route) in map form is a problem for cartographers because a straight line drawn between them on most map projections does not represent the shortest distance. Not knowing this fact often results in misleading map graphics — for example, maps showing the range of missiles from a launching station. To get round the problem often requires the use of map projection software which allows the cartographer to centre an equidistant map projection on the point of origin of the route.

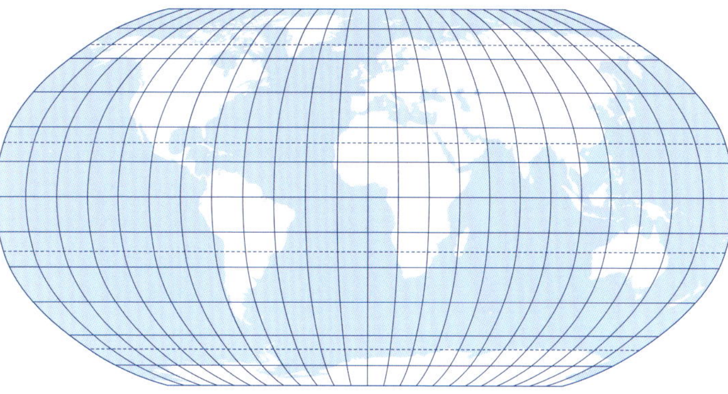

Although neither equal area nor conformal, the Robinson projection is popular because shapes 'look' right.

### did you know...?

Web Mercator is used by many popular web mapping applications. It is a variant of Mercator calculated using spherical development of ellipsoidal coordinates. It is not quite conformal and rhumb lines are not straight lines. The benefit is that the spherical form is much simpler to calculate.

Web Mercator, or WGS84/Pseudo-Mercator, is neither strictly ellipsoidal nor strictly spherical. Since distortion is so great towards the poles it is cut off at 85°N and 85°S. For fast delivery of online mapping the sacrifices are deemed acceptable — in local areas distortion is not so significant.

A GREAT DEAL OF MAP INFORMATION is held and used in GIS format. A geographical information system is a computer-based means of storing both the graphic information about map objects (the points, lines and areas) and their attribute information (names, object types and other information). A GIS is more than just a digital map because it has a linked database of information about features on a map. The two main data models used in GIS are *raster* (based on a scan of a map) and *vector* (where objects are digitised as points, lines and areas).

GIS consist of a software program and data. The software programs are capable of capturing, storing and analysing any spatially related data. The data in a GIS will often include topographic mapping acquired from a national mapping agency or a commercial cartographer which provides context for additional, specialist information, such as local authority or population census data. GIS information is *georeferenced*, meaning that it is related to a fixed coordinate system such as the National Grid.

GIS information is structured in layers. Basic topographic map information is itself divided into different layers: major roads, minor roads, rivers, contours, boundary lines, buildings (sub-divided into many different types), etc. Additional layers of information show specific themes or subjects of relevance to the user: location of water pipes, data on health or property values, and 'listed' buildings are examples.

GIS can also combine map information with aerial photographs or satellite images, addresses, postcodes, census data, market research information and much more.

### did you know...?

GIS software and principles underlie the customised mapping and Earth-viewing websites you're probably familiar with. If you type in a postcode to find a map of an area, it's a GIS that links the postcode to the right bit of map.

Extract from Europe map in its finished form, ready to print (below) and the same area in vector outline (right).

Courtesy: Global Mapping

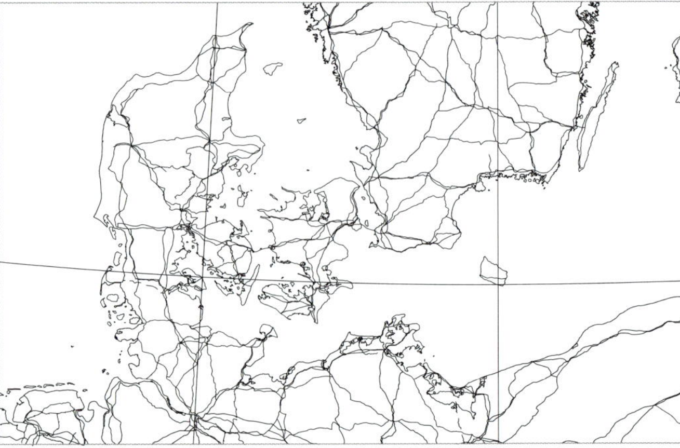

The advantage GIS have over paper maps is that, as well as storing maps in a digital form, they can be used to analyse that information — e.g. 'find me all locations more than 5km from a fire station'. They can also model landscapes in 3-D, perform statistical analysis, find shortest routes between a series of points, and allow measurements (such as the calculation of areas or distances) to be easily obtained.

Typical output from a GIS query is a map. Often the map will include all or some of the topographic map forming a base, with the thematic information shown on top of it. The map-creation tools in GIS software have improved hugely, allowing the cartographer more flexibility and creativity in designing good maps. However, when default settings are used, the results can be very poor maps that do not communicate a message well. Typical problems include poor colour choices, too many data classes, legends poorly laid out and organised, poor type placement, and little attention given to titles, margins and the other items that make a good map graphic.

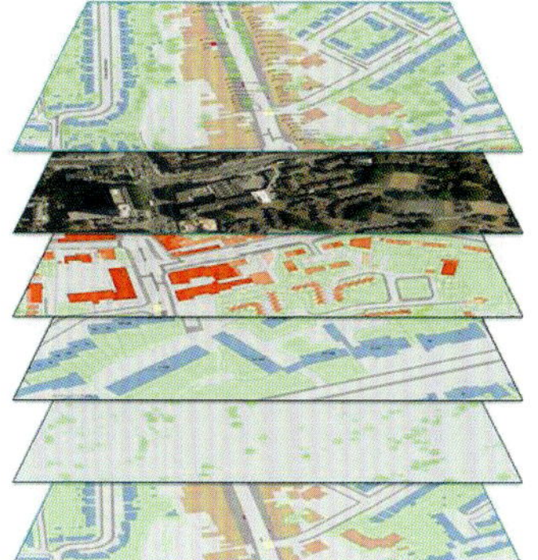

Example of GIS layers for UKMap: A database of feature-rich mapping across Greater London.
© 2017 Insurance Services Offices, Inc.

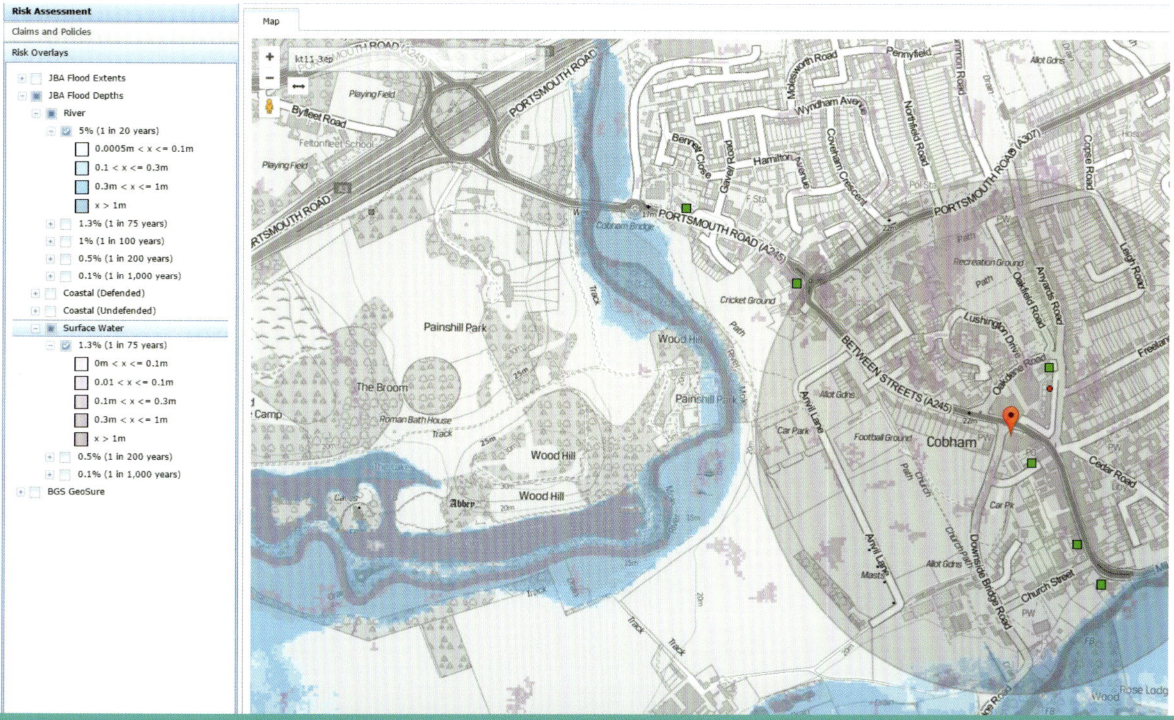

Risk Insight — precision risk assessment (flood) GIS using OS base mapping and BGS data.
© Europa Technologies Reproduced with permission of the British Geological Survey © NERC © Crown Copyright and database right 2017

**A**LTHOUGH MAPS VARY GREATLY in content and appearance, the way of producing them often follows a similar path. Thought should go into the planning and design of maps, as well as into their production.




### Defining the map

Understanding the reason for making the map is the first step to take and will help to determine the parameters of design and production. Having a clear understanding of what the map is aiming to achieve helps the design process. Even where the purpose of the map is outlined for you (e.g. it will form part of a local plan), considering the target audience, where it will appear, the budget and production time will help to define the variables that you *do* have control over: things like size, layout, colours and choice and size of symbols.

### Design

To produce a good map, the design process is critical. Although it is easy to design 'on the fly' while creating a map on computer screen, taking a little time to create rough layouts, sketches and trying alternatives

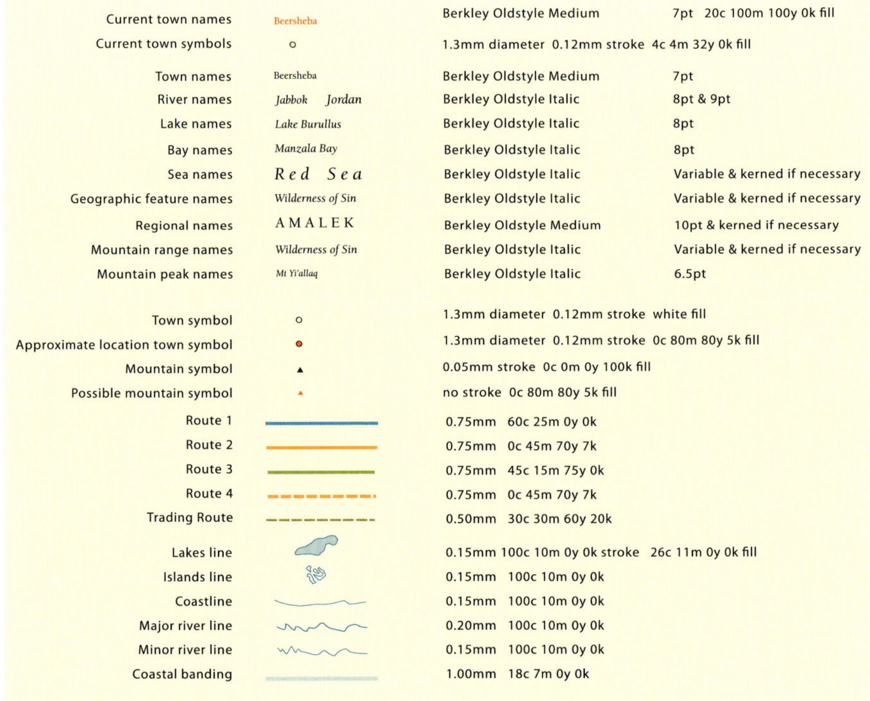



A typical design brief with specifications for linework, colours, symbols and fonts.
Courtesy: Millennium House, Australia

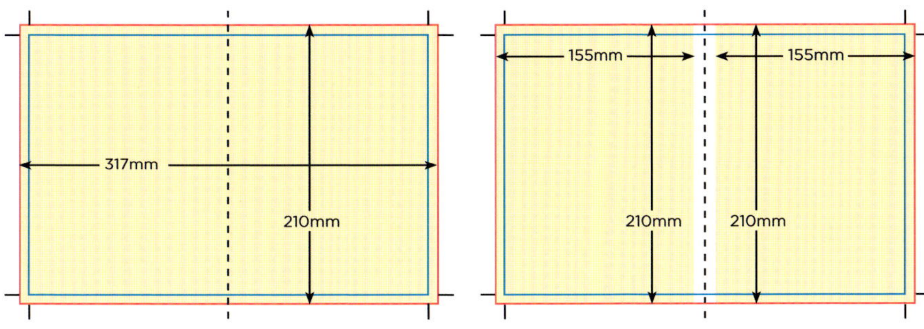

Page dimensions
and layout for atlas
production.

can lead to a better final design and quicker production. As well as the map itself, titles, legends and scale bars and other marginalia have to be accommodated and they should be considered as part of the overall design. If it's for printing, decide on the page format: landscape, portrait or square. Then do some brainstorming of graphic ideas, for example by sketching 'thumbnails' of how the elements might be laid out, using page outlines of the right proportion. Simply looking at similar existing maps and analysing what works and what could be improved is a useful step to take. Creation of a proper specification sheet is very helpful for quick reference and ensures that the map is consistent.

## Selecting data from existing maps

Many new maps are derived from existing maps, and it is rare treat to create a map from brand-new information. Creation of your map might therefore involve the selection of features from an existing map which will form the base, editing it and generalising if necessary. Don't assume that you have to show all the information on the base map; it is easy to clutter the map unnecessarily and a base map needs only show enough information to give your thematic data context.

Historic mapping can be scanned and georeferenced within a GIS snd either used as a backdrop or as a basis for digitising detail and adding current data.
Courtesy: LovellJohns

## Map production

Production usually takes place using either graphic design software or the map-output tools of a GIS. If the information you need to map is not already digitised, a careful hand-drawn compilation is scanned and matched to the base-map data, and

the linework and other features are traced. New information is added and the selected themes symbolised. The right colours and design styles are applied as the map is made — e.g. line weights, symbol sizes and type styles are all assigned. In reality, a design may change as the map is created and it becomes obvious in the light of experience that improvements can be made.

## Adding the non-map material

The map itself is usually only part of the overall graphic. At a later stage, the marginalia are added, including the map's neatline and border, title, legend, scale bar, reference grid or graticule, location maps, other inset maps, and often copyright notes, acknowledgements and data source information. Watermarks may also be added for copyright protection.

OPPOSITE:

Map production

Courtesy: Harvey Maps

An illustration of a map with all its marginalia and the extra graphics that can be added from the GIS data, all presented in a neat and aesthetically pleasing layout.

© 2017 Insurance Services Offices, Inc.

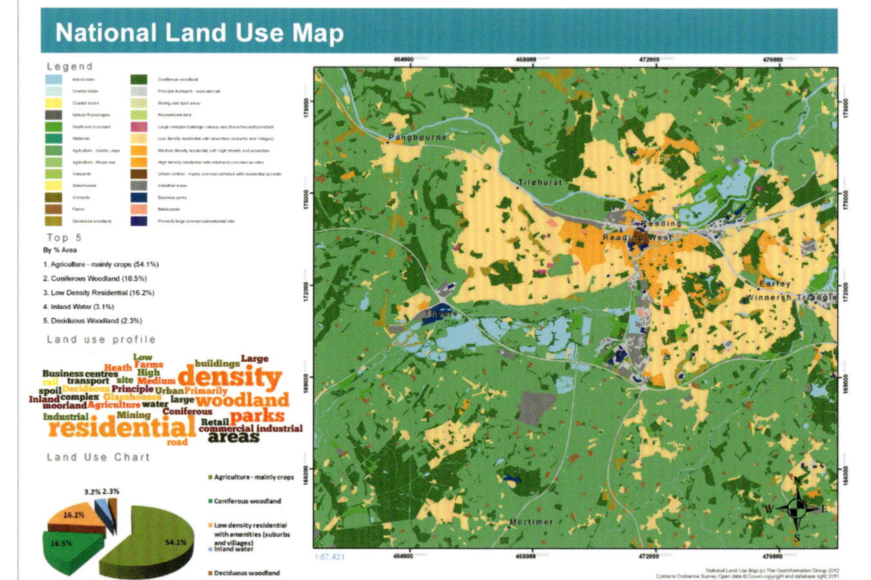

## Checking and proofing

Maps are checked for errors as you go along. Basic printed proofs are very useful as it is often easier to spot errors on a printed map than it is on-screen, even if it is only intended for electronic delivery. It can be really helpful to ask someone else to check it, with a fresh pair of eyes.

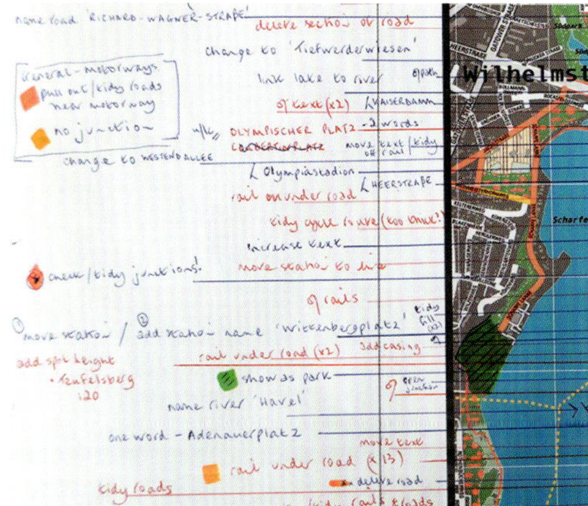

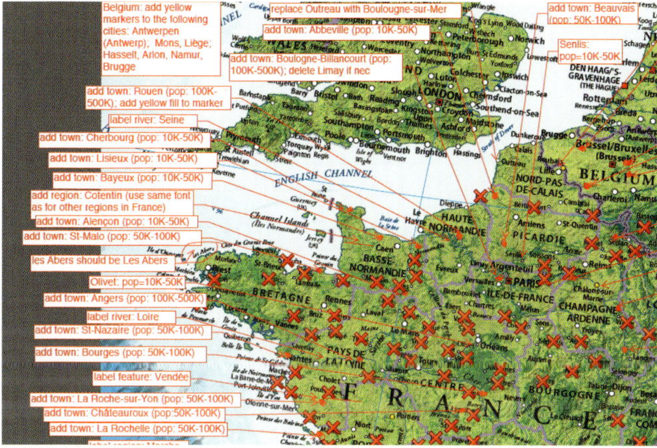

## Output

Maps for screen use remain in digital form only but they usually have to be written into a file format suitable for web or other platform delivery. It is worthwhile testing that they show on a web page or mobile screen as you intended. Printed maps are usually printed in four process colours (cyan, magenta, yellow and black) which combine to produce the full range of colours. Spot colours are sometimes introduced to match corporate colour schemes and some products are designed to be printed in only one or two colours. The normal output required for printing is a press-quality PDF. Registration marks, trim marks and colour bands may be added.

Proof corrections marked on a paper proof (left)
Courtesy: The Future Mapping Company

and proof corrections marked on a PDF (above).
Courtesy: Millennium House, Australia

Large-format lithographic press for high quality printing.
Courtesy: The Future Mapping Company

GOOD USE OF COLOUR on maps can greatly aid the communication of mapped detail. Colour can also enhance legibility and contrast and it can make the important elements of the map stand out from background material. Most map designers are working with colour, and colour output is the norm for both printed and on-screen maps.

Although many maps are now delivered to their readers in both printed and screen format, designing maps for the two media requires a different approach.

### Printed maps

Maps are printed using the four 'process' colours: cyan, magenta, yellow and black (CMYK). Cyan, magenta, yellow and black are the primary hues used to mix colours with ink. They equate with artists' paints in that when you mix them together, the results gradually get darker. Although cyan, magenta and yellow mixed together in equal measure produce black, the density of the overprinted colours is not sufficient for lettering or fine linework and so a separate black ink is printed.

Cyan, magenta and yellow are called the subtractive primaries because the addition of more colour (e.g. the ink) subtracts the amount of light reflected from the surface (e.g. the paper).

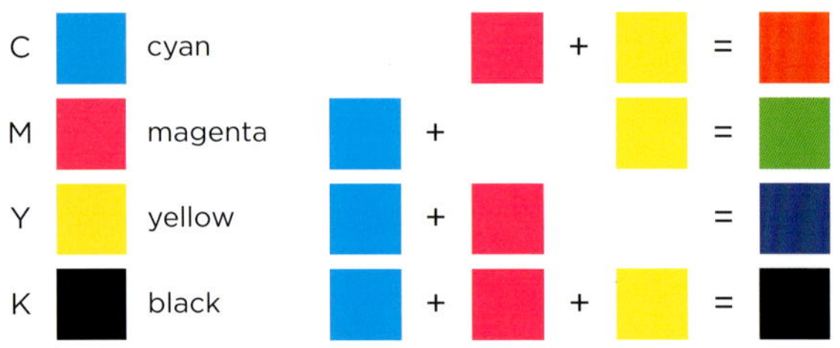



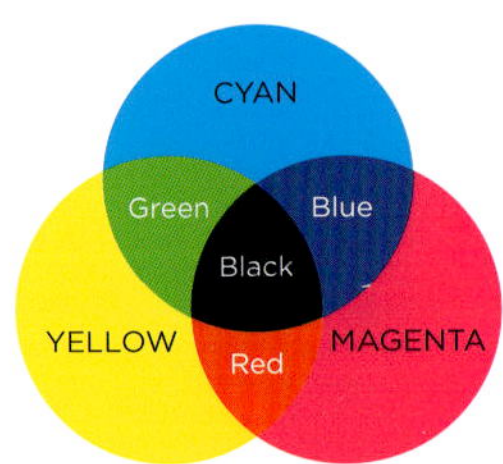



Different percentages of C, M, Y and K produce thousands of other colours. The proportions vary from 0 to 100% ('solid') of each. The printing process begins by the generation of colour separations to obtain primary screens.

Each screen consists of tiny dots — the higher the percentage, the bigger the dot. The screens are printed on top of each other to form the colours. When viewed on the printed page, the dots optically merge to produce the final colour. The process of separating each colour for the four screens for printing is done automatically when output is to a PDF.

The khaki colour comprises 40% cyan, 20% magenta, 50% yellow and 10% black. The separated screens are shown here enlarged to 20 dpi to illustrate the relative dot sizes and screen angles.

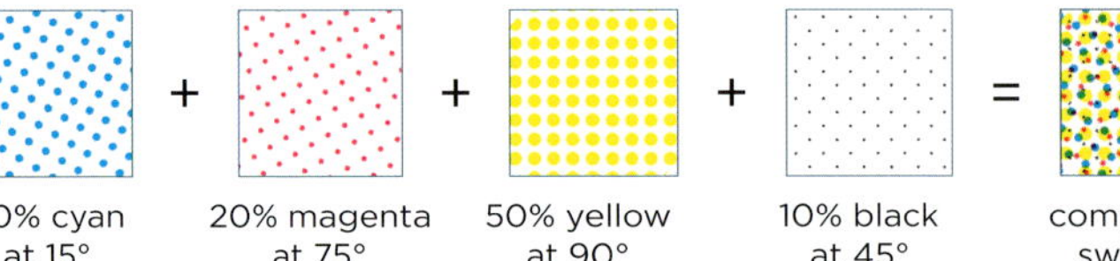

If a particular colour is required (e.g. for a company logo) it can be printed as a special colour (a 'spot' colour) in addition to the process colours, but would add to the expense. The colour is often specified as a Pantone® ink colour and can be used in the map design too.

There is usually a mismatch between the colours on a computer screen and the final colours when printed. If you always use the same printer, use a printer's colour chart to find the colour shades you want. The colour chart shows C, M, Y and K mixed in different proportions, usually in 5 or 10% increments.

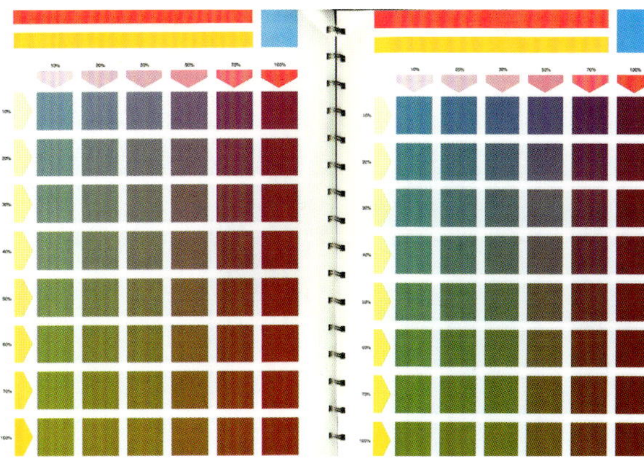

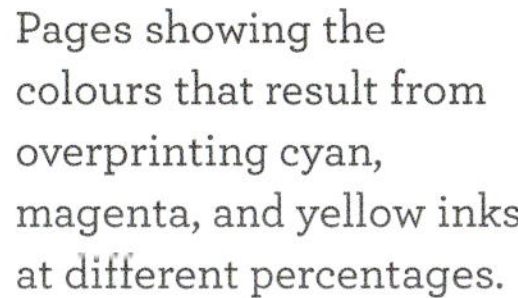

Pages showing the colours that result from overprinting cyan, magenta, and yellow inks at different percentages.

OPPOSITE:

**Choosing the colours.**
Courtesy: The Future Mapping Company

## Maps on screen

Images on television, computer or telephone screens are created by mixing red, green and blue light (RGB). When C, M and Y are added they get darker, but when RGB is added together, they get lighter. Equal quantities of R, G and B produce white on the computer screen.

As with process colours, many more colours can be made by mixing RGB in different proportions but in reality fewer colours can be differentiated on-screen than can be seen on paper. Screens are made up of pixels (picture elements) and the smallest symbols must be larger than the pixels that display them. As screen resolution has increased, this issue has become less important.

More information on designing for screen output is given in chapter 22.

**did you know...?**
The Pantone® Matching System (PMS) is a standardized colour reproduction system of spot colours which are created from 13 base pigments plus black.

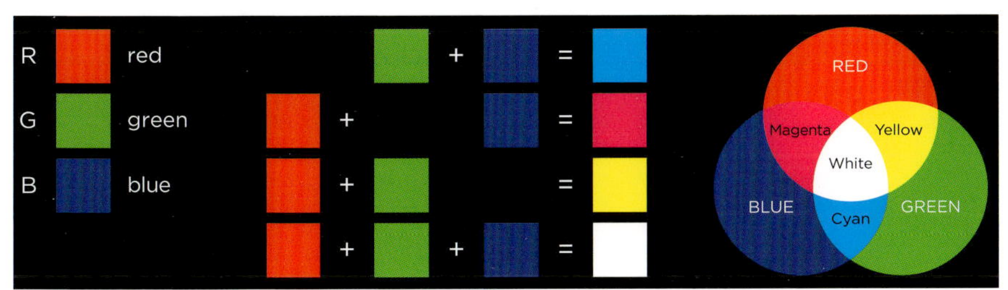


Red, green and blue are called the additive primaries because colour is being added to a black background.

**G**OOD CARTOGRAPHY IS IMPORTANT because a well-designed map communicates its message better than a badly designed one.

If you look at the many maps you encounter every day — printed or on screen — you'll soon realise that their quality varies widely. Some maps are clear, attractive graphics which are easy to read and convey their message well. Some are adequate but not inspiring, while others miss the point completely and fail to communicate at all.

Good design is also important because, like it or not, we tend to judge the quality of the contents by the quality of a map's design. It is unfortunately the case that a map which has good data behind it but is poorly designed is seen as being less valuable than a map with poor data behind it but is well designed. The reader quite reasonably confuses the quality of the message with the quality of the messenger.

It's often hard to know exactly why one map works well while another doesn't. However, as the visualisation expert Edward R. Tufte put it, 'graphical excellence consists of complex ideas communicated with clarity, precision and efficiency.' Maps which don't work are often unclear, imprecise and inefficient.

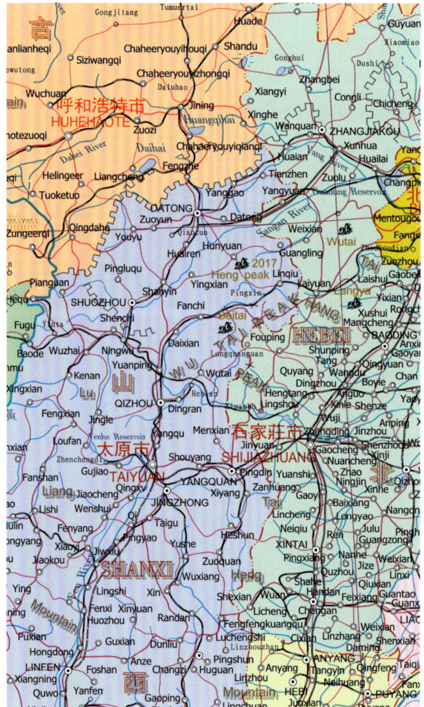

China wall map: A totally new design was required here as well as a major editorial revision and update of the content.
Courtesy: Global Mapping

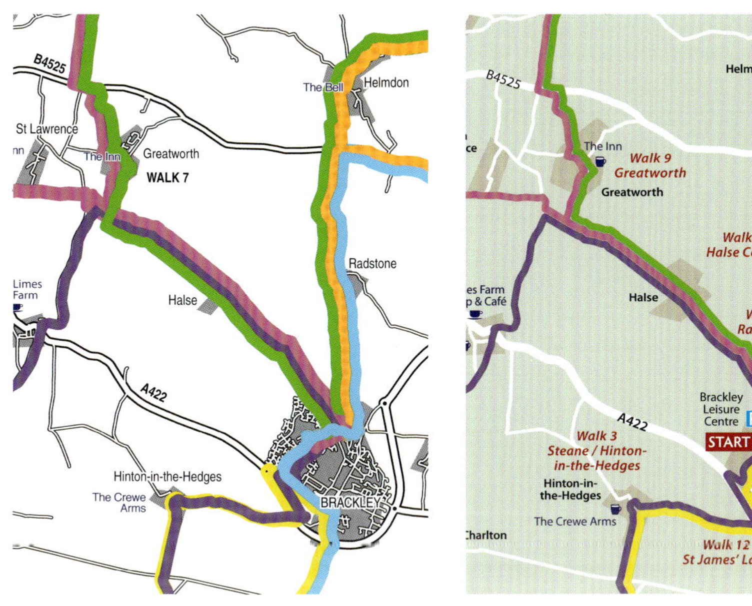



Cycle Routes: Using the same data, improving the design and adding useful information:

— change base mapping to coloured background with white roads

— remove minor roads, especially in town centre

— tidy the cycle routes ensuring they are all the same width and offset equally from each other

— add the start point for the cycle routes at the leisure centre where there is parking

— add nearby walks.

Courtesy: Global Mapping

There isn't a neat formula for making the perfect map. A map is more than the sum of its symbols. Although computers are involved in all stages of map production, they cannot be programmed to design excellent maps, and in fact many maps produced in a GIS using its default settings are poor graphics. Good map design benefits from a bit of artistic talent, a sense of harmony, of form and of the appropriate use of colour. However, by understanding the different elements that go into making a map, and reviewing design options, a cartographer can turn a poor map into a good map by applying a few basic principles.

Services map: Redrawn using accurate base data and with improved symbology.

Courtesy: Global Mapping

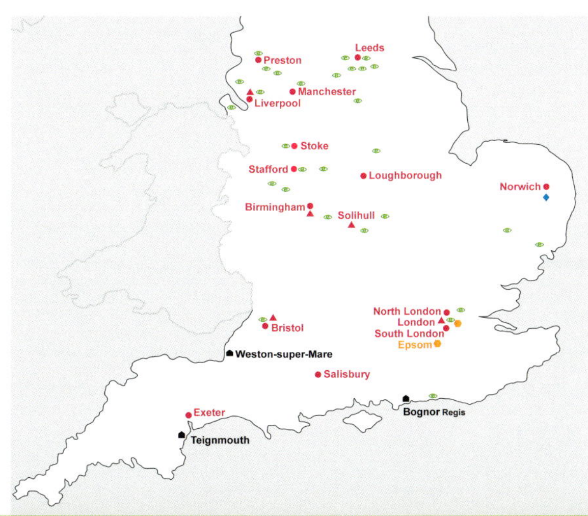



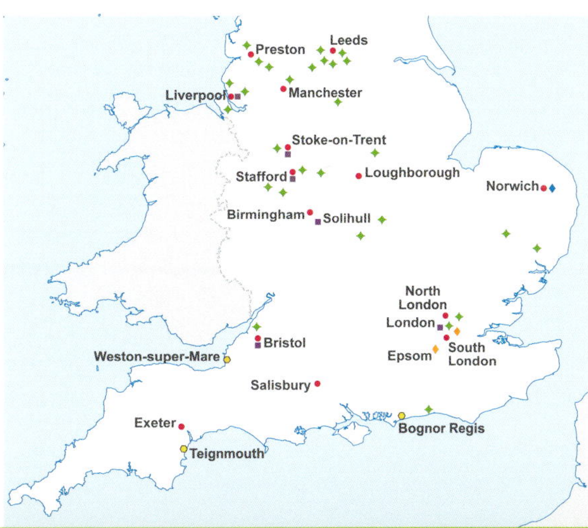

A MAP CAN ONLY BE as good as the information that goes into it. Working with source data of high quality is important if a good, reliable and honest map is to be produced. Digital map data vary in quality and, if you can, it's worth finding out something about the quality of source data. Check the metadata:

- Input scale — if the digital data have been captured from a map at a smaller scale, it will have been generalised, and you can't get out of it more than went into it. Generalisation removes information that can't be put back by 'zooming' into it. Cartographers always compile a new map from the larger to the smaller scale so that they can generalise appropriately and consistently.

- Date of capture or update — information can go out of date surprisingly quickly.

- Map projection — vital for matching maps of large areas of the world, and important even for smaller areas where data come from different sources, especially over long periods of time when standard projections may have changed.

> **useful tip**
>
> Test the quality of data by careful checking against a source you know to be reliable or against an area you are familiar with.

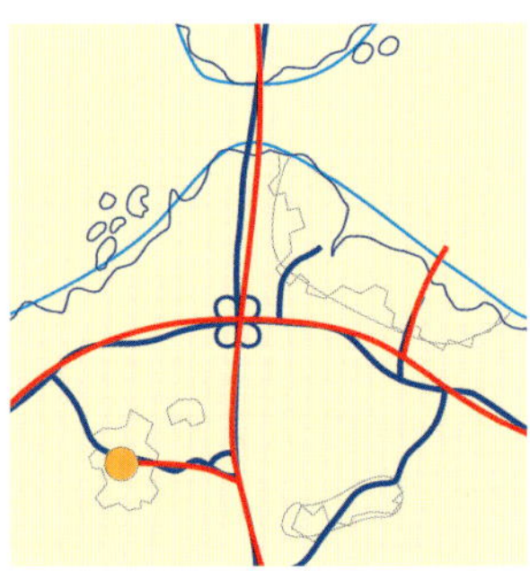



Both datasets overlain to illustrate the differences between data captured at large and small scales.

If you're combining information from a number of different sources, you need to assess each source to see if each is of the same or a comparable standard. Data quality is hard to define consistently, but a data source should be 'fit for use'. It's worth considering whether your data match up to four criteria: completeness, consistency, accuracy, and similarity of scale.

- Completeness: Do the datasets show all the features they should? For example, have some major roads been shown, but others omitted?

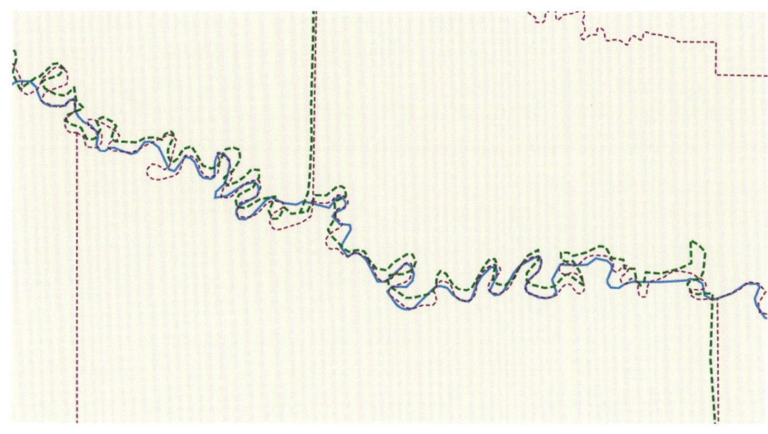

An example of mismatched data where the state boundaries (green) and the county boundaries (purple) have been taken from different datasets that have been captured at different levels of detail. The two sets of boundary lines are not correctly aligned to each other and, in addition, neither set perfectly matches the rivers (blue).

● Consistency: Do the datasets show similar objects in the same way, and have they been classified consistently? For instance there may be confusion between land use (e g recreation) and land cover (a grass field). Roads may be classified as primary, major and minor, but the definition may vary from one dataset to another.

● Accuracy and precision: Are the objects in the right place, and is the level of detail used appropriate for the scale? Digital geospatial information (like printed maps) cannot just be a miniature representation of reality, so you need to assess if the source data is reasonable in its generalisation. Compare a printout of the data with a map you already trust. This can show whether objects are in the right place, and whether source datasets are complete.

● Source scale: Remember that you can get more information from a large-scale map and so you need to compile information from larger to a smaller scale map. Ideally, all sources of data should have been compiled at about the same scale otherwise you will have a lot of work in making them compatible.

Different dates and different projections for source data results in this mismatch of Oxfordshire county boundary alignments.

HONESTY IS VITAL in cartography. Maps are accepted as believable documents, and they are widely trusted. But mapped information may assume an authority which it does not merit. Moreover, as cartographers often derive a new map from an existing one, mistakes on one map have a habit of appearing on other.

It is usually impossible for the map user to distinguish between the quality of the information and the quality of the presentation. If they are both well designed and produced, a map with good information and one with bad information are equally credible. It's only when you come to use it that you find the errors. Most errors are unintentionally introduced, but sometimes they are deliberate. For example, the use of green as an area fill colour may have associations with being environmentally friendly, and can be consciously used to persuade a map reader that a development proposal is more acceptable than it might seem if red were used.

**did you know...?**

Mapmakers can introduce small errors on their maps as copyright traps to find out if others are copying their products illegally, e.g. small roads apparently under construction, or a stream that doesn't exist.

Compare two official maps (above) with the Communicarta map (right) which has been prepared following extensive field research. Note particularly the careful use of interchange/connecting symbols and the clear relationships between all the stations on the Communicarta map. All the stations are represented in their correct relationships with each other, correctly named, and in relatively correct geographic positions.  The names in boxes indicate terminus stations, a detail missing from the official maps.  Black keylines indicate interchange links to adjacent stations by way of walking at street level, again not shown on the maps above.

Courtesy: Communicarta

The process of generalisation deliberately introduces errors because it is essential to simplify the varied, real world to depict it on a graphic. We eliminate unnecessary information, simplify complex geography, and exaggerate the size of some features to make them visible because they

Christopher Saxton 1579

John Speed 1611

Joan Blaeu 1645

Robert Morden 1695

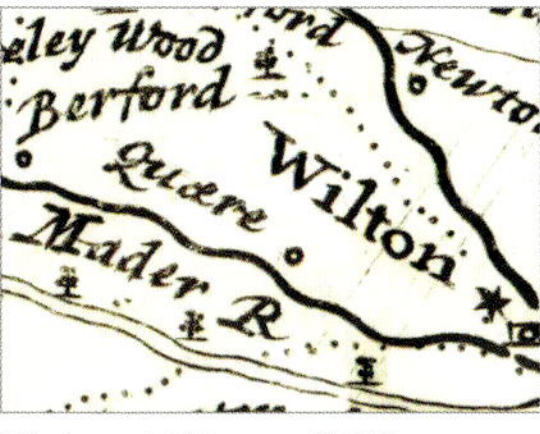
Richard Blome 1716

Hermon Moll 1742

Thomas Kitchin 1760

are important  Map users tolerate a lot of distortions and errors in maps, for good reason: it makes them more useful and understandable.

The idea of an 'accurate' map is one which is hard to define, but users have a feeling for what is accurate and what isn't. Accuracy Includes:

- positional accuracy — objects are in the right place
- completeness — objects are on the map when you expect them to be
- consistency — similar features are treated in the same way.

All three of these attributes are measurable, and national mapping organisations have statistical standards to which their maps should conform — for example, that all contours are accurate to within half the contour interval. However, a lot of map information comes from sources which, reasonably, may not have the same standards.

Conscientious mapmakers have a duty to minimise potential errors on a map, and try to see how a map might mislead. This is especially the case in statistical mapping which usually involves the classification of data (explained in chapter 29). Since there is no foolproof way of classifying data, care has to be taken to show the statistics in an informative and honest way.

Maps gain, not lose, credibility if you add notes about their reliability. So, 'positions approximate', 'scale varies across map', or 'selected roads only' are useful notes, not confessions of poor quality. Stating the source of data and the method of classification is also a good way to inform the map reader.

A reasonable question to ask yourself when making a map is: 'would I trust my own map?' If so, then the map reader will too.

In 1579 Christopher Saxton showed a village to the west of Salisbury, but without a name.  In 1611 John Speed used Saxton's map as a source for his own.  Seeing an unidentified village, he noted against it 'quaere' ('investigate' in Latin) as a reminder to himself to check on the name of the village. Unfortunately, his engraver Jocodus Hondius thought that this was the name of the village and added it to the map.  Thereafter, as one cartographer copied another, the name was perpetuated for 150 years until Thomas Kitchin's New English Atlas of 1760 showed the village as North Burcombe.

Courtesy: John Leighfield



If you label something that needs to be verified later, then do it in bright magenta, for example, so that it doesn't get missed.

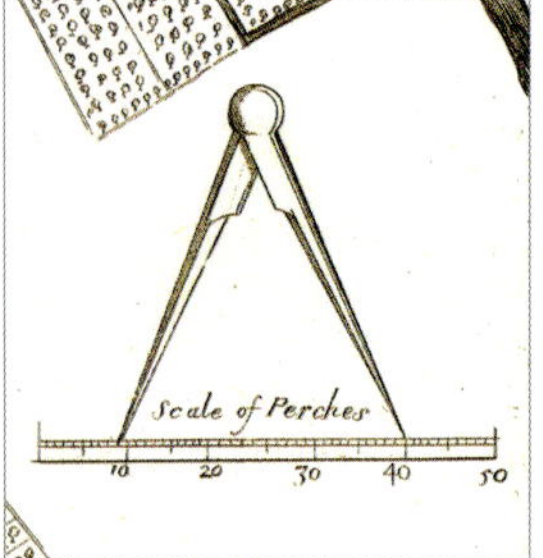


ALL MAPS ARE REDUCED VERSIONS of the world, or scaled down. Scale is expressed as a ratio of the map distance to the ground distance — a scale of 1: 250 means that one map unit represents 250 of the same units on the ground.

The most usual ways of showing scale on a map are:

- as a statement in words — one inch to one mile, two cms to two kms

- as a ratio — 1:10,000

- as a representative fraction — $\frac{1}{1250}$

- or graphically using a scale bar —

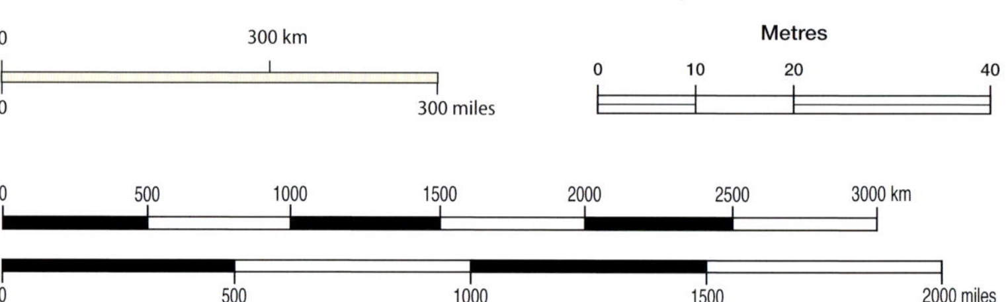




Confused by small and large scales?  Small-scale maps require a small sheet of paper to show an area with a small amount of detail; large-scale needs a large sheet for the same area and shows large amounts of detail.

The advantage of a graphical scale bar over the other methods is that it will remain correct if the map is altered in size (e.g. on a computer screen). All the other methods become wrong.

Scale bars will often carry both metric and imperial units since a significant number of map readers will prefer one system over the other. Metric measurements are often put on top, but reverse that convention if the metric units extend further than the imperial so that the scale bar looks more grounded as a result.

Often scale bars on topographic maps have a section to the left of zero which subdivides the major units that are shown to the right.



If a map carries a 'not to scale' note it probably means that the scale of the map is variable, e.g. larger scale in the centre to show more detail and progressively smaller scale towards the edges of the map.

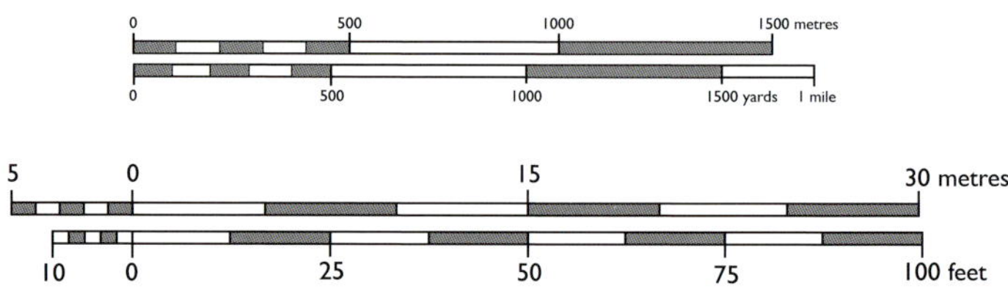

Not all maps require a scale bar. If the map shows a very familiar area or is a sketch-map, then a scale may not be necessary or appropriate. But if the map is to carry any authority, then it probably needs one. Small scale thematic maps may not include a scale bar.

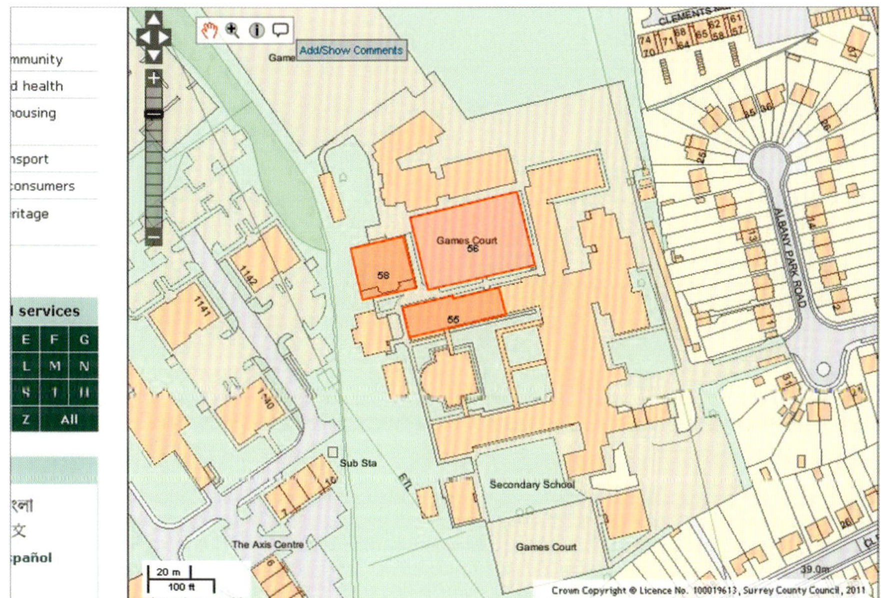


GIS web mapping: A simple scale bar that changes with zoom level and remains visible in the corner of the image.
Courtesy: LovellJohns

Care has to be taken on maps of the world or large regions because the deformation introduced in creating the map projection means that the scale bar will only be correct for one line on the map (very often the equator on a world map). Any other line will not strictly be correct because scale varies across the map. It is helpful to put an explanation such as 'Scale at the equator' or 'Scale at 40° North' next to the scale bar. However, for small areas (such as town plans), the scale will hold good across the whole of the map.

A variable scale bar such as seen on a world map illustrating the change of scale with latitude.

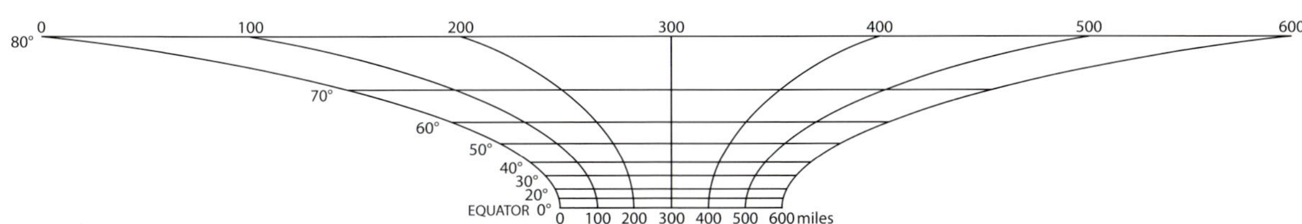

SINCE MAPS ARE SCALED REPRESENTATIONS of the world, they can't show every feature of the landscape and should never try to. The smaller the scale of the map, the more selective it has to be in what it can show. For instance a general reference atlas will show more detail than a school atlas for the same area.

Generalisation is the process of simplifying the complexity of the real world so that it becomes graphically clear at reduced scale, and puts due prominence on the features that should be seen. Feature selection is the first step in that process — deciding what elements are to be included on the map. At its simplest, for example, this means a navigation chart for small craft does not require a detailed road network on land. Conversely, a road map will not show the essential information that a yachtsman needs to navigate safely.

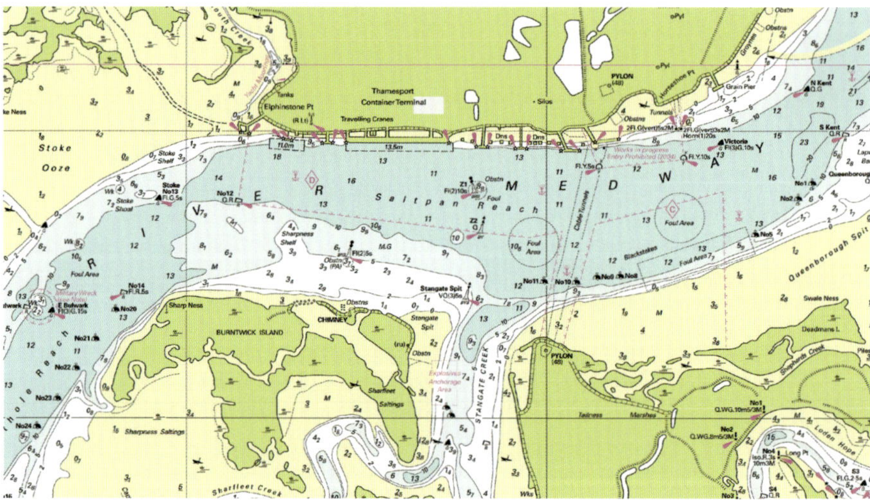

Navigation chart for small craft with detail on the land limited to features that can be seen from the sea — landmarks such as tall buildings and masts.

© Imray Laurie Norie and Wilson

The content of a map is governed by a number of factors:
- the scale of the map
- who the map is for
- what the map is for
- what the map has to communicate
- the conditions under which the map will be viewed and
- the medium of delivery.

In addition, the map literacy of the audience can be a major factor in the process. For example, a thematic map depicting complex statistics will be easy to interpret by someone who knows the subject whereas a layman may struggle to understand what is being protrayed.

Once the content and purpose of the map is ascertained, the selection of information is a straightforward issue of deciding whether an element is essential to the understanding of the map. Anything more than this will result in unnecessary clutter and confuse the message. When compiling a map the greater task is often deciding what to leave out rather than what to include.

An example of far too many roads for the scale and purpose of the map. There needs to be much sifting of that element.
Courtesy: Global Mapping

Geographic data usually come in distinct categories — for example, roads comprising motorways, primary routes, A roads, B roads, etc. As scale reduces it is not simply a case of omitting the bottom categories since an unclassified road in a rural area will have greater local significance than an A road in a dense urban area where even primary routes may need to be sifted. A great deal of editorial work must be carried out to create a meaningful small-scale map from complex datasets.

Large-scale maps and plans contain more detail than small-scale maps and so, if you are reducing the scale of a map you almost always need to generalise. Cartographers aim to work from large-scale source information to smaller-scale final output. Generalisation means losing detail, and it's a one-way process. If you increase a small-scale map in size you don't get more detail, only over-generalised symbols. To compile a new map, you should always go from large-scale to small-scale.

UK wall maps: Editorial selection of content as scale reduces from A0 size (left), through A1 (centre) to A2 (right). Note the progressive sifting of information within each element and the removal of Welsh name forms and counties on the A1 and A2 maps.
© Tiger Moon

# Generalisation and feature selection



Once the information has been selected and the scale of the map is known, the generalisation process can begin. This involves modifying the data so that the map can convey just the right amount of information at an appropriate level of detail to enable accurate geographical interpretation. A number of different techniques are employed to achieve this aim.

- Simplification: Making features easier to read by leaving out small details and variations — e.g. smoothing a meandering river. Simplification also involves the elimination of minor features.

- Exaggeration: Small or narrow features can be increased in size if they are locally important or display a significant characteristic. Roads show up on a map by making them wider than their true scale.

- Displacement: The separation of symbols may be required for clarity, and is often necessary as a result of exaggeration.

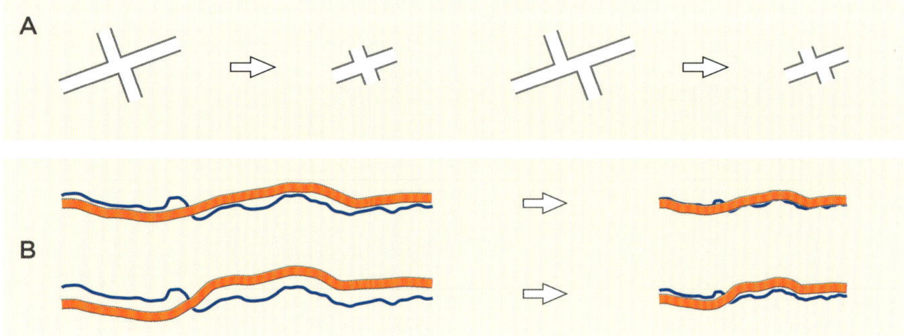



- Merging: Joining adjacent objects together — e.g. merging individual buildings to form a built-up area. This process also includes typification

**useful tip**

Think of generalising a line as drawing over it with a very thick pen — the kinks and wobbles are naturally smoothed out.

Simplification: The important features of the line have been retained.

Exaggeration: Part of a map feature may be small but significant in retaining its character.

Displacement:

In example A, a staggered junction has been displaced to make it more visible.

In example B, the road is displaced to prevent it overlapping with the river.

which means taking the broad characteristics of an object and including them in the symbol, but removing minor features.

Merging: Joining objects with the same symbolisation.

● Classification: The amalgamation of similar geographical features which would be differentiated on a larger-scale map. For example, roads may be classified as motorway, primary, A and B on one map, but grouped together as 'Roads' on a smaller-scale map. Land labelled as growing different crops on one map may be amalgamated as 'cropland' on another map.

Classification: Grouping similar features into a new symbolisation.

Generalisation also involves changing the form of symbols as the map scale reduces. For example, complex lines can be simplified and geographical areas depicted as point symbols.

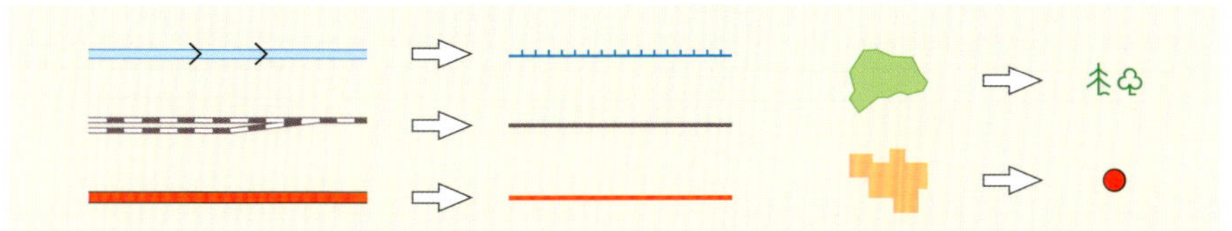

The amount of generalisation and its nature is dependent on the scale of the map, but also on the importance of one object in its locality. A single small building may be eliminated in a map of an urban area but, on a map of a remote highland area, plotting that building could be of great value to the map reader, so it would be retained.

Some maps do not work well because they are over-generalised. For example, features on a town plan such as large buildings whose planform could be shown are depicted as point features. This makes judging the scale of the map harder to assess by eye, and groups together places that can be shown as areas with features that really do merit point symbols.

**useful tip**
Take care if using automated generalisation software — human judgement often gives more pleasing results.

Reivers Cycle Route:
A variety of different
symbols used.

© Stirling Surveys

The graphic variables
that can be employed
for qualitative and
quantitative mapping.

A MAP IS A GRAPHIC which shows geography in the form of symbols. On large-scale maps and plans (larger than about 1:10,000), the symbols are fewer, simpler and objects can be shown on or close to their correct positions. Smaller-scale maps need to simplify the complex world more and carry more symbols, and the location of objects is more generalised. Simple lines can show a church's outline on a plan, but a point symbol is used on a small-scale map.

Most maps use three types of graphic symbol to show almost all features:
- points — for example a well, or a town on a small-scale map
- lines — for example a railway, or a national border
- areas —for example a lake, or a county.
    Map text is also a symbol, and its presence assigns specific identity to the other symbols.

Points, lines and areas are varied in order to communicate different types of map information. The graphic variables include:
- size
- shape
- colour
- lightness
- orientation and
- pattern or texture.
    These can be employed in isolation or as a combination of different effects depending on the type of map — topographic maps will require a totally different set of symbols to a thematic map. Map readers naturally understand the variation in symbol forms and assign meaning to them but a comprehensive legend should explain unfamiliar elements.

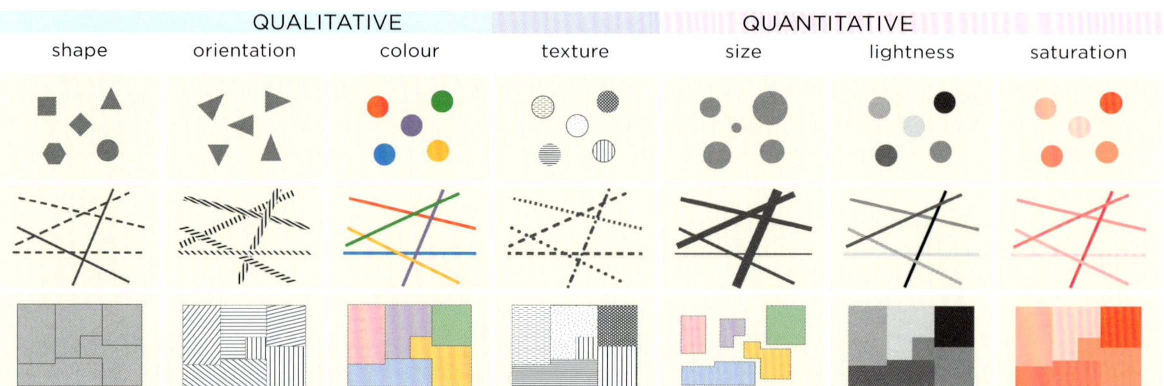

## Points

Point symbols are mostly:

- geometric — based on circles, squares and triangles
- conventional — like a red cross for a hospital, or
- mimetic — looking like some aspect of the object they're representing, such as an aeroplane to represent an airport.

A simple filled square can represent a town on a small-scale map, but add a cross to it and it represents a church.

geometric          conventional          mimetic

## Lines

Lines can be simple strokes, but most are more complex. For instance, roads are often cased with two lines enclosing a coloured fill. Many lines are varied by adding to them, such as the conventional symbol for a railway, or broken up (*pecked* or *diced*) like many boundary lines. Ticks can be added on one side of the line to denote a canal, for example.

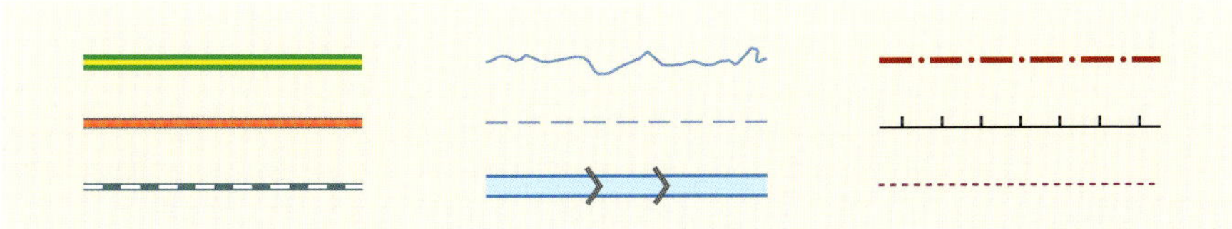

## Areas

Areas form the background to a map and may be important symbols in their own rights. They are either filled with colours or with textures and patterns. On complex maps such as geology and land use, the addition of letters or symbols to the areas gives valuable visual clues to make identification much easier than a vast range of marginally different colours. Areas often have a thin bounding line or *keyline*.

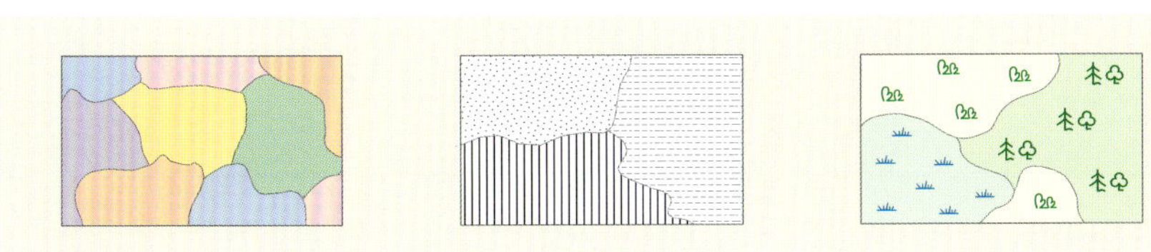



Consider whether you actually need a printed keyline to separate areas of colour. If you do, keep them as fine lines and think about their colour — solid black is seldom a good choice. A broken line can be an option.

John Speed, Lancashire 1610 — sugar loaf hills.

FROM EARLY TIMES, maps have tried to show the ups and downs of the landscape. Early maps used pictorial symbols of hills to depict relief, but it is hard to tell much about the heights of the mountains or hills shown or about more gently changing scenery in between. Depicting relief has always been a challenge for cartographers, and showing height and slope without graphically overloading the map can be difficult.

There are a number of techniques employed to depict both height and slope on maps.

## Spot heights

Spot heights show the height at points above (or below) a datum, usually sea level.

## Contours

**useful tip**

Contour values should read uphill and only the index contours need to be labelled.

Contours are lines of equal height printed in a neutral colour such as brown or orange. They are usually derived from aerial photography. The contour interval will vary with the scale of the map — the smaller the scale, the greater is the contour interval — but also with the terrain, and flat areas may have an increased contour interval to show locally significant variations in height. Maps will often show spot heights at summits or low points in addition to contours. On a maritime chart, contours relate to the chart datum and the spot heights shown are depth soundings. Although widely used as a method of relief depiction, many people find it hard to visualise a 3-D landscape from contours.

## Hachures

Hachures are patterns of fine lines which run parallel to the direction of slope. Line length varies according to the length of slope they depict, and steeper slopes are represented by thicker lines or closer spacing.

Old and new: Hachures on OS 1:2500, Yorkshire [West Riding, York] 1st edition, sheet CLXXIV.10, 1892 (right) and on Historical Map of York 2014 (far right).

© The Historic Towns Trust

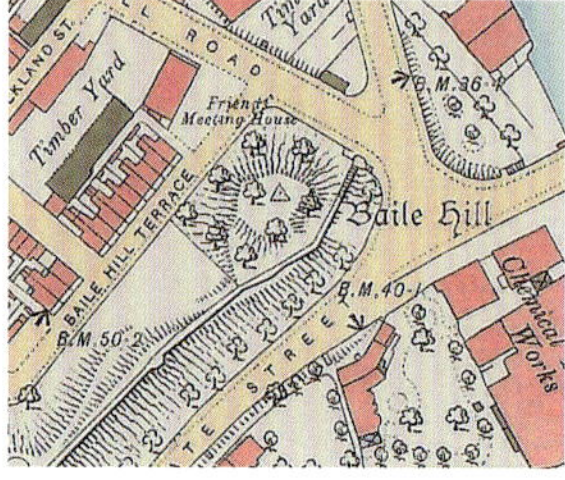

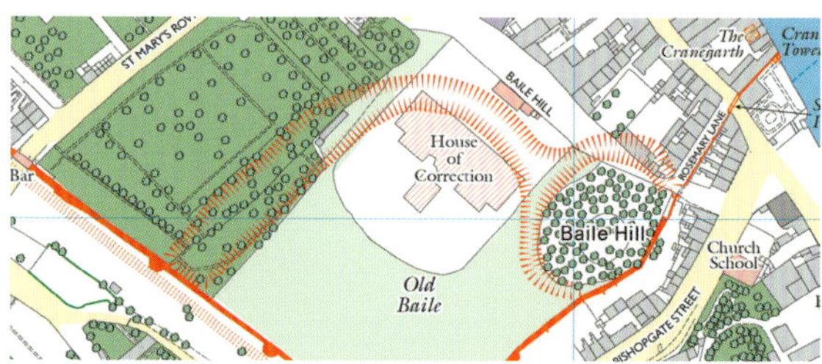

## Hypsometric tinting

Heights are grouped into different bands and a colour applied to each band, typically ranging from greens and yellows for low areas, through oranges to browns for the highest areas.

## Hill shading

Hills and mountains are made to stand out by illuminating them, making slopes in one direction appear sunlit and slopes in the opposite direction appear in shadow. Hill-shaded maps almost always show the light coming from the top left (usually NW — a direction the sun rarely shines from). If it comes from the bottom right, mountains look like valleys and vice versa.

Many maps use more than one method to show relief, for instance by combining spot heights, contours and hypsometric tinting.

Digital methods of depicting relief use height data captured from maps to model a landscape in a GIS. They include digital terrain models (DTM) and digital surface models (DSM). A DSM is similar to a DTM but includes the surface material not included on a DTM, incorporating images from aerial photography, satellites or LIDAR draped onto the DTM.

Lake District map showing different techniques for depicting relief and terrain.

© Harvey Maps

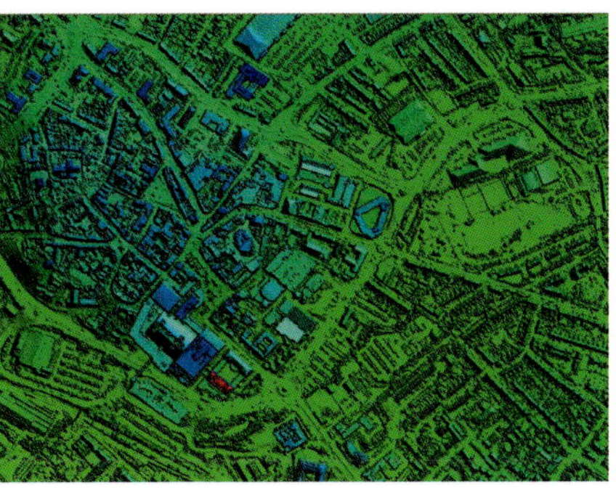

Aylesbury: DTM (far left) and DSM (left).

© 2017 Insurance Services Offices, Inc.

A WELL-DESIGNED MAP communicates its message much better than a badly designed one. There isn't a formula for making a well-designed map, but there are key issues to consider. Many of the following principles are explained in detail in later parts of the book.

### Clarity and legibility

Good maps work because you can easily read them, without confusion, without difficulty, and intuitively understand them. Maps may fail because the graphic is too complicated (trying to show too much information) or by using symbols that are indistinguishable, or set too close. If the map looks overloaded, consider removing some of the information if you can. Ensure that all symbols are large enough to be seen, and distinct from one another (separated by at least 0.2mm on a printed map).

### Hierarchy and structure

Group map information into broad classes (e.g. settlements or rivers), then subdivide by size or importance to reflect the ranking that the map reader expects. Vary the size of symbols and text to reflect relative importance: larger and darker usually means more important. One object imposed on top of another (a road crossing a river, for instance) establishes hierarchy.

### Colour and pattern

Choosing appropriate colours is essential for a good map. Standard colour schemes are used on series such as geology maps where subclasses are shown in groups of similar colours. Conventional colours (like blue for water, green for vegetation) are common but colour balance on the map may suggest alternatives. Avoid complex patterns (particularly overlapping ones) that give you visual indigestion.

> **useful tip**
>
> Concentrate on making the most important element (the subject of the map) the first thing to be noticed. Then structure the other information around it.

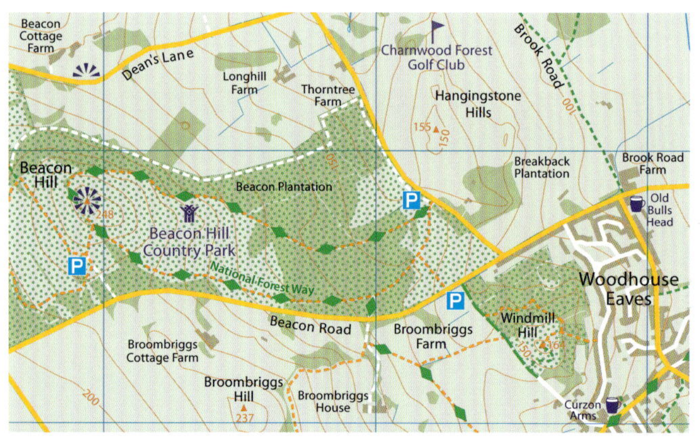

Visitors' Map of Charnwood Forest Regional Park: A complex set of data structured so that the information reveals itself in the correct order according to importance. Note the subtle use of thin, coloured halos in the underlying colours which allows the features of interest (purple), contour values (orange) and long distance path names (green) to sit above the green spot pattern of the open access areas.

© Global Mapping

## Visual contrast

Visual contrast comes through varying symbol size, colour, shape and orientation. Ensure that symbols are sufficiently different to be correctly identified — for instance that a series of circles varies enough in diameter to be seen without confusion.

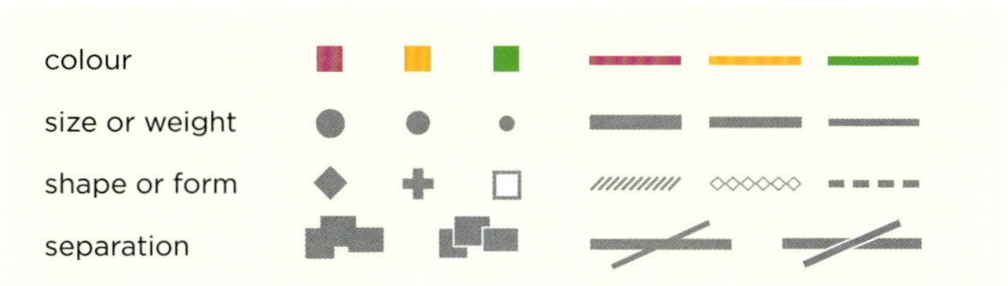


Contrast can be created by varying a symbol's characteristics.

## Figure and ground

The figure-ground relationship between map symbols requires determining what is more important (the *figural*, on which your eye settles) and separating it from the background (the *ground*). Symbols can be made more figural by having fewer of them. Closed forms (like islands or country outlines) stand out, as do dark objects and familiar objects (like your own country's shape). Small objects stand out from larger backgrounds. Some colour combinations, like yellow on black, are naturally figural.

Poor contrast between land and sea.

Contrast improved by bringing the most important elements to the foreground.

## Balance

A map needs to be balanced, both internally and as part of the page layout. Although you cannot change the geography of the area shown, you can alter the non-map elements (title, legend, insets and illustrations, for example) to complement it. All map elements should be looked at together to achieve balance.

## Good typography

Good typography makes a map; bad typography can ruin it. Avoid fonts which are too large or fancy and keep type sizes in proportion to the map. Titles and text which are too big look unprofessional and distracting. For computer mapping, special fonts are used since many traditional fonts don't transfer well from paper to screen.

UK Low-flying Chart:
Specialist mapping
with many complex and
unguessable symbols
that are familiar to
airline pilots but not to
the layman, even with the
help of a comprehensive
legend.

© No. 1 Aeronautical
Information Documents
Unit

THE BEST MAPS WORK BECAUSE they're well designed and the mapmaker has thought about the purpose of the map and the target audience. Even when the map brief is given to you, there are many design variables which are in your hands, and spending a little time considering the characteristics of map readers may help in the design process.

The term 'fit for purpose' is often used by cartographers and what it means is that the map should show what it is required to show in the clearest way possible. Unfortunately, perfectly good maps created for one use can instantly become poor maps when they find their way into the hands of a different user because the design and content are inappropriate for that use.

Think about why the map is being made:

- what's the reason for making the map?
- what information is it communicating?
- where will it appear, at what size, and in what format?

Think also about your audience and their characteristics. Are they:

- already knowledgeable about the subject?
- learning about it?
- novices?
- the 'general public'?

The purpose of the map and the nature of its target audience should affect your design. A knowledgeable audience will accept a sophisticated map, with more complex symbols, less legend information or explanation. Many geology maps and maritime charts, for example, are highly complex, technical documents requiring a high level of interpretative skill from their specialist readers. In contrast, road atlases are usually widely understood, and don't require reference to a legend for basic features.

Many maps will, of course, have a wide readership, varying in map-reading ability, and you can't satisfy everyone. It is interesting to note that the amount of time in formal education devoted to the interpretation of graphics is far less than the time devoted to the study of language and mathematics, the other primary means of communication in our society. As a result, mapmakers have to be more attuned to the likely skill levels of their readers. Perhaps the target audience of the 'general public' can be defined as having normal vision and average intelligence.

Tailor your map to the level of the audience by varying the number of categories of information, and choosing graphic symbols which vary in complexity according to the skill level of the reader. If the target audience turns out to be extremely varied, aim your map at the most important readers, or the largest likely group of readers.

It is important to consider the medium of presentation too, before making your map. The design of maps prepared for use on a computer screen is very different from those intended for eventual printing.

Three maps from school atlases with increasing levels of complexity appropriate to different age groups.

Junior School Atlas.
© Philip's

Foundation Atlas.
© Philip's

World Atlas.
© Philip's

There are many things to consider before you can start creating your map and there is a vast array of different map types. However, all maps are best constructed in the order: areas, lines, points, text.

Area symbols are drawn first (**1**), line symbols added on top (**2**), then points (**3**), and finally text (**4**). Each symbol layer will obscure the layer beneath it, but map readers are skilled at unconsciously joining up objects which are separated by an overlying symbol. For example, when a contour value is added, the contour line is supressed where the number is placed, but we understand the line to be continuous.

Text sits on the highest visual level and is added last. Text is allowed to obscure other symbols, as long as there is enough of the symbol visible for the reader to make sense of it. Text is often surrounded by a halo or hold-out (**5**) of a light colour (white, or the same colour as the underlying area) to make it stand out from the background information.

Choosing the right symbol is a combination of understanding what feature you are mapping and a sense of good graphic design.

Europe Postcodes: Base detail shown in grey with postcode data in colour bands per country sitting well above the background.

© Global Mapping and XYZ Maps

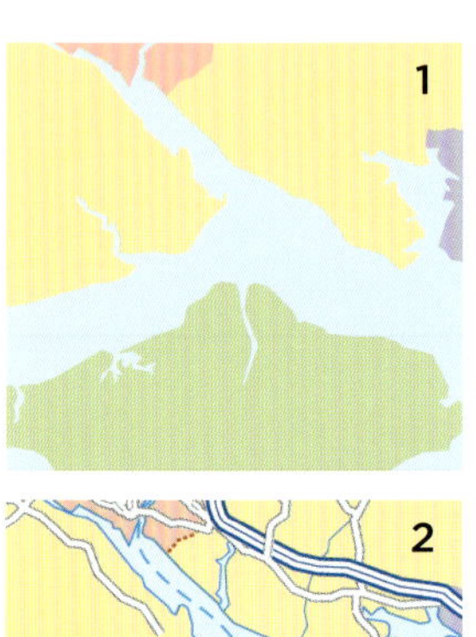

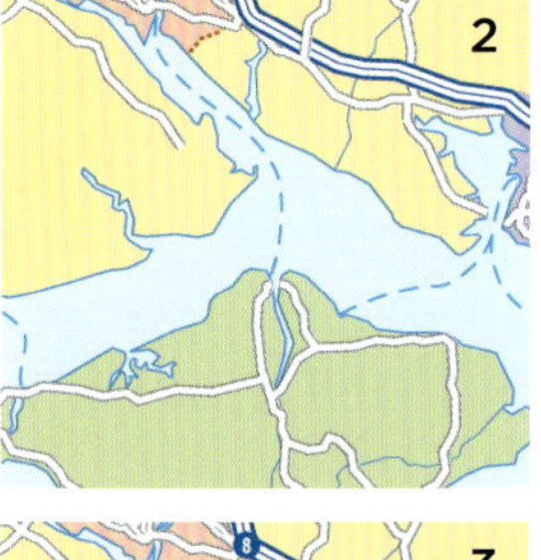

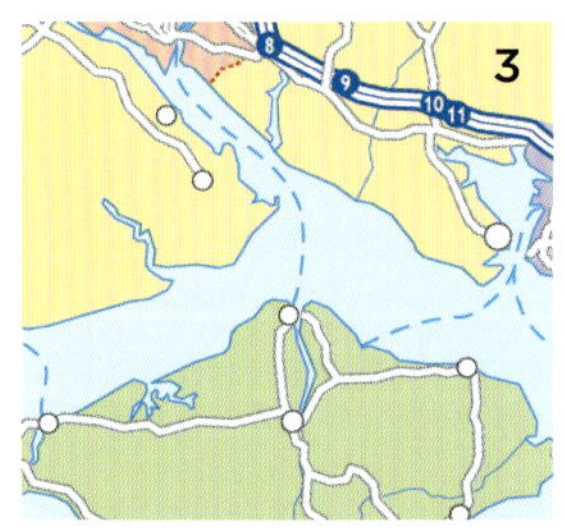

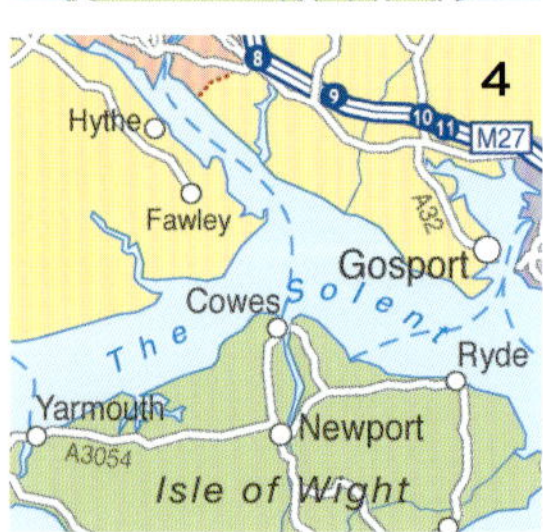

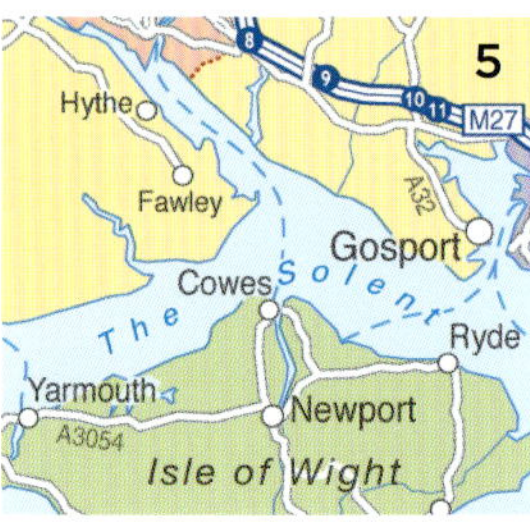

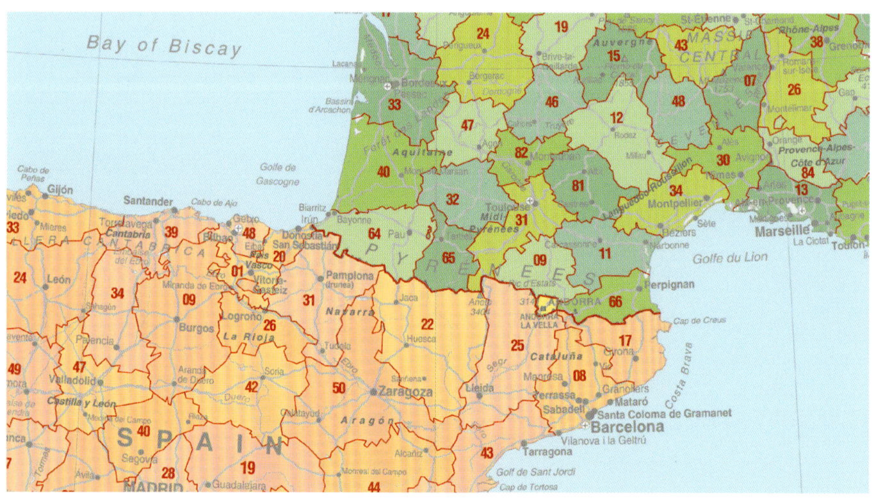

Firstly, decide if your map information is qualitative or quantitative. If it's qualitative, then all features are of equal importance. On a land-use map, for example, all classes of land use are of equal value, and therefore you should try not to make one class look more prominent than another. The symbols' shape, colour and design are varied, but symbols should all appear to be of equal importance. Symbols for qualitative maps are explored further in chapter 33.



World Continents and Regions: No one region dominates the map — text style is used to differentiate between continent and large country names.
Courtesy: Tiger Moon

If the map shows a quantity or 'ordered' data (most statistical data, or different classes of road, for instance), then the size, colour range or colour intensity of symbols is deliberately varied to make one feature look more important than another. Quantitative maps are explained more in chapters 30 and 34.

Then choose a symbol which relates to the map's size. All symbols must be easily distinguishable, so they need to be legible and vary enough in shape, size and colour to avoid confusion. Which symbol you choose depends partly on the type of feature and partly on the scale of the map. A town will appear as lines and areas on a plan, as an area symbol (where its boundary is shown) on a medium-scale map, but as a point on a small-scale map.

The Dynamic World: Earthquakes magnitude 6 – 7.4 and >7.5
The white outlines enable the clear depiction of so many overlapping symbols whilst lifting them above the Natural Earth background.
The combination of larger size and darker green aids the visibility of the >7.5 magnitude symbols.

Courtesy: Global Mapping

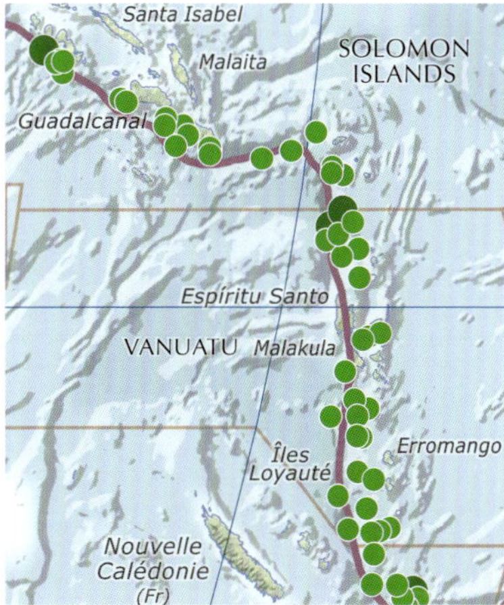

### Choose the right colours

● Important features need to stand out from the background and colour choice can help.

● Where appropriate, use conventional colours (green for vegetation, blue for water, for example) and associative colours (like blue for cool, red for warm).

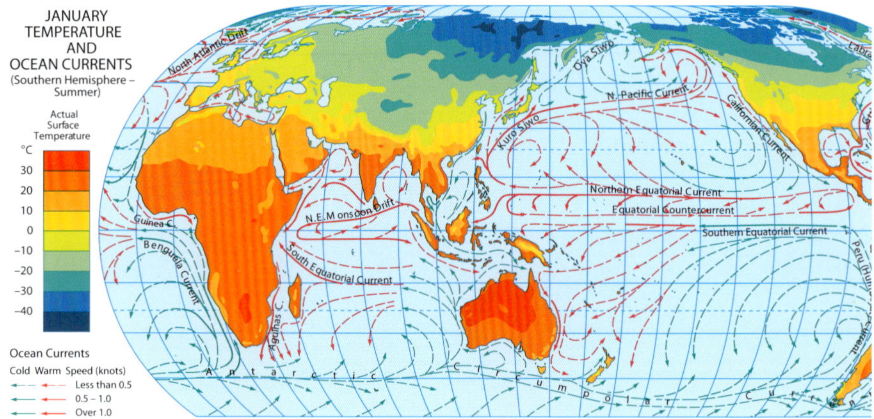


Australian Edition of The Philip's Modern School Atlas: Appropriate colour association for depicting temperatures.

© Philip's

● Colours look different depending on their backgrounds, so don't choose them in isolation. The perception of colour also alters with the size of the coloured symbol. Large areas of dark or saturated colours can look unintentionally figural.

Colour interactions: The orange swatches are identical in colour but don't appear so. The red swatch appears to advance in front of the blue — note how the yellow advances most.

● Areas surrounded by lighter colours appear darker and areas on a darker background appear lighter.

● Good maps often use subtle, balanced colours, and bold colours can look inappropriate. However, a flat range of colours can look too bland.

### Qualitative maps

● These are maps where one symbol is not meant to be more important than another, just different — ecology and soils maps, for example. Don't show a hierarchy where one is not required; choose colours which have the same visual value and look equally important. Strong, saturated colours will stand out, so reserve them for small areas to make them show.

Spectral colours

● Using part of the spectrum can make a good sequence.

## Quantitative maps

● Quantitative maps should show a hierarchy. Darker colours have greater importance, so the best sequences of colour are from light to dark to represent 'least' to 'most'.

● There are a number of different colour sequences which can be used:

— progression of a single colour, e.g. light green to dark green

— progression of two or three colours

— mixing grey with a single colour, and reducing the amount of grey.

● More information about the use of colour on choropleth maps is given in chapter 30.

Single colour progression

Two colour progression

Three colour progression

Value progression

## Some other helpful pointers

● Small letters and small symbols need more intense colours to show.

Adding a fine keyline enhances the visual contrast of pale colours.

● The most legible text colours are black, dark brown or blue on white.

● The greatest contrast is between yellow and black.

● Background colour affects the legibility of coloured type.

Background colour affects the appearance of symbols too.

● If you're limited to black and white, then use no more than six shades of grey for fills, and don't use a grey that's more than 70% black, especially if you want to place type on top of them.

| 0% | 3% | 15% | 26% | 41% | 55% | 68% | 85% | 100% |

● If you use more than 12 different colours on a map, it becomes very hard to interpret.

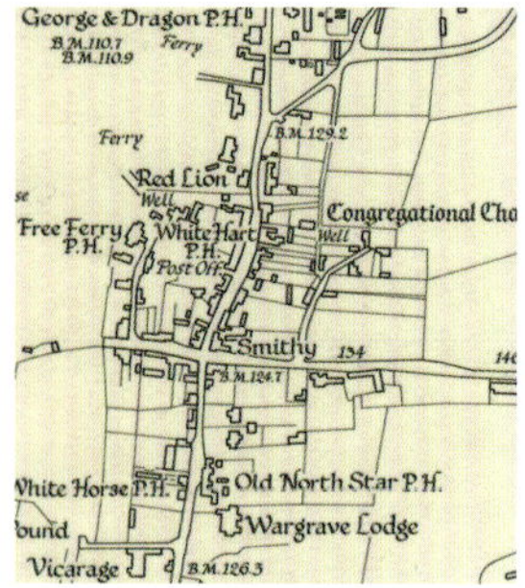

Beautiful hand lettering for a traditional look.

© Phoenix Mapping

GOOD TEXT CAN ADD MUCH to a map; poor text detracts from an otherwise good map by making it difficult to read.

There are six aspects of type which can be varied:

- typeface (font or fount)

  Arial  Helvetica  LITHOS  Myriad  Skia  TRAJAN  Verdana

- type style, determined by the choice of font

  serif:  Baskerville  Optima  Palatino  Times  Warnock
  sans serif:  Century  **Futura**  Gill  Gotham  Helvetica  Trebuchet

- type characteristics

  UPPER CASE  lower case  SMALL CAPS  roman  *italic*  **bold**  **semi-bold**

- type size, usually expressed in points (72 to one inch)

  5pt  6pt  7pt  8pt  9pt  10pt  12pt  **16pt**  **20pt**

- character spacing (or tracking) and line spacing (leading)

  l e t t e r   s p a c e d
  **extended**
  condensed

  Newcastle
  upon Tyne
  9pt type on 12pt leading

  Newcastle
  upon Tyne
  9pt type on 10pt leading

- colour — by associating a colour with a feature class.

  *River Thames*  **Regents Park**  **C I T Y**  **Tower of London**



## Typefaces

Simple and popular typefaces work best on maps and shouldn't cause problems to readers of digital maps. Special fonts have been created that allow greater on-screen legibility — for example, Arial, Helvetica, Verdana, Geneva, Trebuchet and Century Gothic have been recognised as popular in web design.

Limiting yourself to a few fonts on a map and varying them by size, characteristic and spacing often works much better than using many fonts. Serif typefaces are often used for natural features and sans-serif for human features, but there are many exceptions to this rule. Serif typefaces aren't old-fashioned, either — they can be easier to read in paragraph texts.

## Type size

Maps use smaller type sizes than you might think, typically 6 to 12 points — most people can read 6pt map text without a problem. Large text takes up space and looks bad. Even map titles look better in small sizes; large map titles look unprofessional and can unbalance a graphic.

Le Jog cycle map: The cycle route is highlighted with overnight stops and distance tables in strong contrast red over the softer colours of the base map. Coloured type in the base map is used to link feature names to linework — road numbers to roads, region names to boundaries, water names to rivers and lochs.
© Global Mapping

Work out a hierarchy of feature class and then vary size, case, spacing or colour of typeface to differentiate them, e.g. distinguish counties from districts by using a larger size for counties. Keep associated features in the same font using size and weight to indicate importance — such as towns ranging from 5pt Helvetica Regular to 12pt Helvetica Bold according to population category.



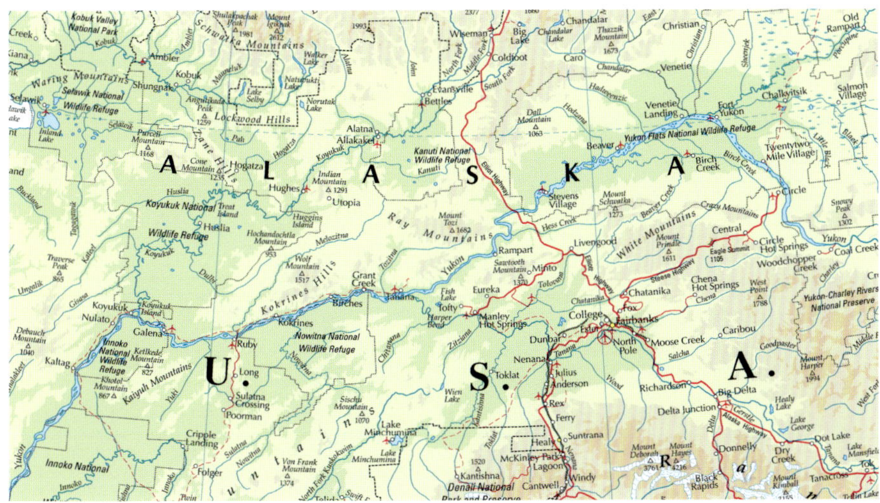

Times Comprehensive Atlas of the World: Appropriate type styles, sizes and fonts with careful name placement applied to general reference mapping.
© CollinsBartholomew

If the type on a map looks cluttered use a semi-condensed variant or 'horizontally scale' it to 80–90%. If paragraph text looks crowded, decreasing the point size but increasing the leading can improve legibility.

## Type placement

For labelling points, there is a preferred order for positioning the label. If you can avoid the label and the point being in line, so much the better to prevent confusion between symbol and label. Keep a consistent distance between points and their labels.

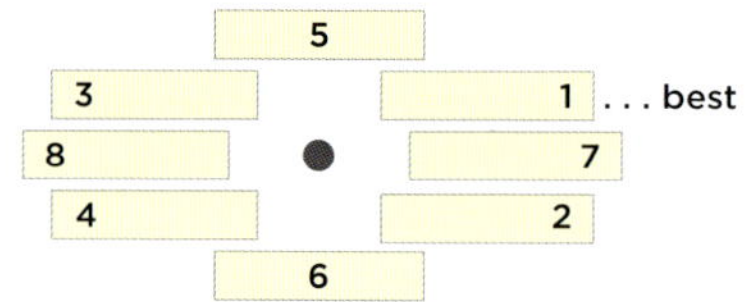


Preferred positions for labelling points.

Romeo    OOSLO

Rome°    °OSLO

Poor alignment of name with town symbol (top) and improved position (bottom).

Automatic placement of town names (left) and improved positions following a custom edit (right).

For labelling lines, position labels above lines where you can, alongside and parallel, but contour or other isolines should be labelled by breaking the line. Don't try to spread out the letters for a long feature; it is better to repeat the label if necessary. If the feature has complex curves (like a river) the label should follow a simpler curve which shows the same trends. Street names shouldn't cross the intersections of other streets and the beginning and ending of a street or part of the street should be apparent from the position of the name.

Type placed along river axes and repeated (right) and curved to follow road alignments (far right).

When labelling areas, spread out names across the extent of the area, but not too close to the bounding edges. Either keep them horizontal, aligned with lines of latitude (for regional or world maps), or make them deliberately curved along the feature — such as a mountain range or valley.

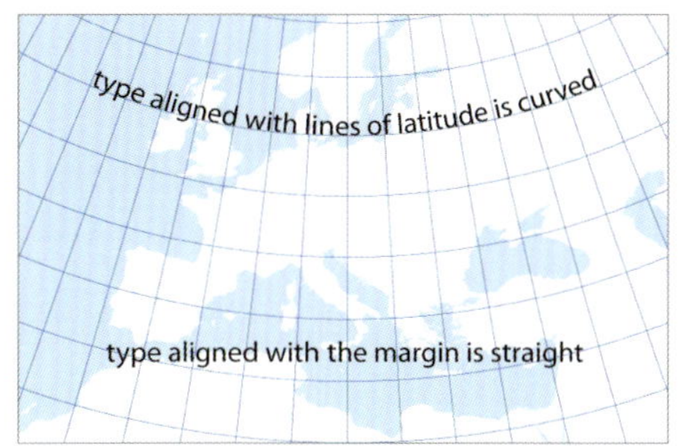


It's fine to space out the letters for an area name, but make sure it still reads as one word and doesn't become a series of disjointed letters separated by surrounding type. If the name needs to spread over a wide area it is best to use upper case since it is easier to read.

Type spaced and curved to fit shape of feature (below left) and varied to suit the size of each area (below).

Type can be made to sit on a lower visual plane by using paler colours or reducing its transparency so that it appears to recede into the background. Thick white type halos can look dreadful but fine coloured halos can work well — see page 48 for an example.

Poor type placement (below left) and improved (below).

THE MAP ITSELF is usually only one element of the graphic you are creating. The other map-related graphics are often referred to as the *marginalia*. The map graphic will usually be framed by a page margin, or by a neatline which denotes the edge of the geographical data on the map. The frame may have essential information within it, such as grid numbers or letters.

A good page layout will balance the map with the marginalia and any text, graphs, photographs and other graphics you need to include, as well as using empty space to good effect. Work out what elements you need to include on the page at the design stage. As well as the map, you'll probably have some of the following:



- title
- map inset
- locator map
- legend
- scale bar or statement
- north arrow
- technical information about the map
- copyright notice
- source acknowledgement
- explanatory texts
- grid numbers or graticule markings
- illustrative graphics — pie charts, etc.

When deciding on the layout, it may be inspirational to use as a model an existing map that works.

- Start with the map itself and consider the shape defined by the geography of the area being mapped.
- Balance the elements (including blank spaces) for harmony.
- Balance the elements around the page's optical centre, which is about 5% of the page height higher than the actual centre.
- Have a bigger margin at the bottom than at the top on portrait pages.
- Elements which are related should be close to one another — e.g. a scale bar or title close to the map it refers to.
- Small, dark objects balance larger, light ones.
- Align page elements vertically and horizontally.

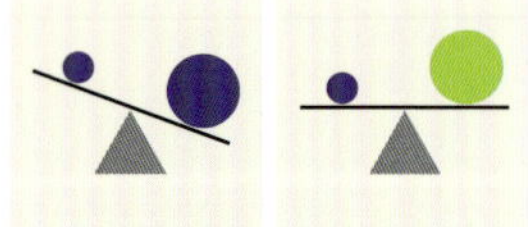

Different colours carry different visual weights.

### Map titles

Map titles should convey the subject of the map, the geographic location and, if necessary, the relevant date. Short titles are almost always better but additional information can be shown as subtitles or text annotations.



**POPULATION DENSITY OF ABERDEENSHIRE**
Adults per square kilometre, 2017
by census enumeration district

The phrase 'A map of...' is not usually required, since it should be evident what the graphic is. Large titles look unprofessional and visually dominant; a title should be proportional to the size of the map and not overpower the sheet. Fancy fonts and text effects such as drop shadows do not generally work well in titles but colour can bring an added visual interest to the map and entice the reader to look further.

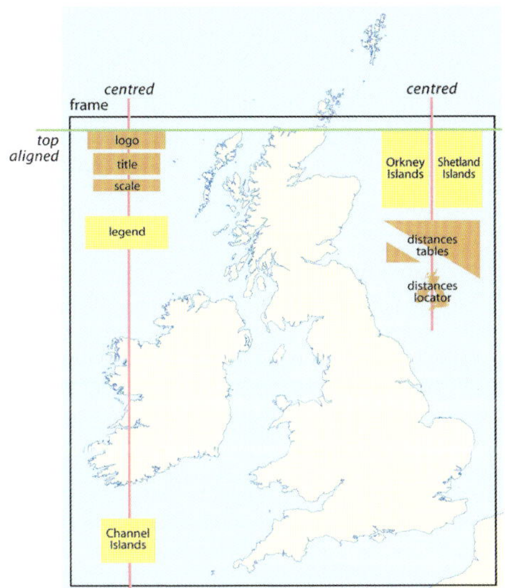


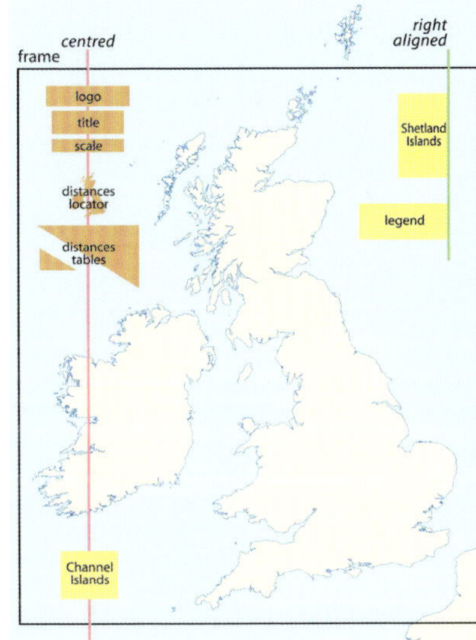


Two options for the positioning of titles and legends — the version on the left shows the most compact layout with both the Orkney and Shetland Islands as insets. On the right the Orkney Islands are included within the frame resulting in a larger map. Geographically, the Channel Islands are too far removed to show in situ so an inset is the only option. Note the symmetrical positioning and edge alignment of the various elements.

## North arrows

North arrows are not always essential, especially if the map is of a familiar area and in the conventional orientation. However, a north arrow should be included if north is not to the top of the map, and on dynamic maps (e.g. maps for hand-held devices), they are very useful if the map changes orientation on-screen. If a north arrow is included, its design and graphic complexity should be appropriate to the general design parameters of the map; unduly detailed or fancy arrows look out of place on a simple map. World maps should have a graticule or a grid on them, rather than a north arrow. For regional or continental maps, including a north arrow is often wrong as north varies in direction across the map and is only correct at the point where it sits.

A north arrow would be pointless on this Sinu-Mollweide projection.
© The Future Mapping Company

## Locator maps and insets

A locator map can be a valuable addition to a map to help the map reader identify its geographical context at a glance.

An inset may be required as an extension of a map at the same scale or to show a particularly detailed area at an enlarged scale. It is important to remember that a larger-scale inset should show more detail (but to a similar specification), not simply be an enlargement of the area from the main map. Showing scale details (a scale bar or statement) is essential if the inset is at a different scale to the main map. If the main map has a grid or the graticule shown it, inset maps should show it as well, with grid numbers or latitude and longitude marked. The area shown in the inset should be indicated on the main map; using narrow-width lines to indicate the inset area helps not to overload the map.

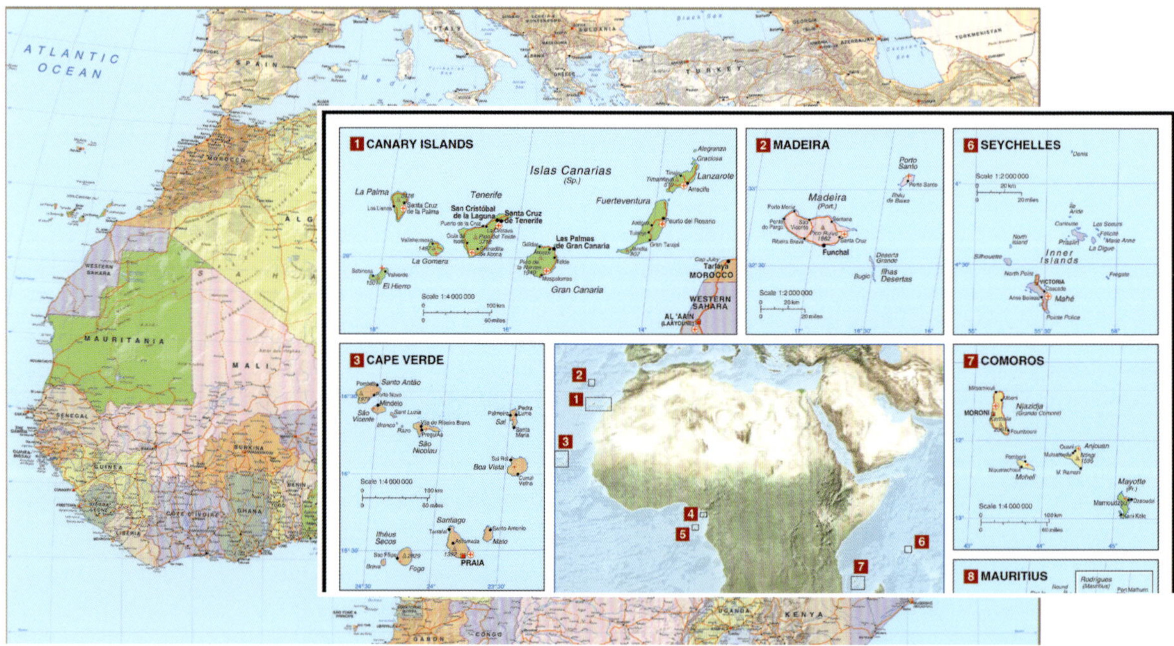

The islands insets on the wall map of Africa are well presented with titles and scale details and their positions indicated on the locator map.  White borders help separate the different maps and the black frame unifies all the insets as a single item. A desirable improvement would be to reproject each inset map with standard parallels and meridians selected to suit the individual inset area, rather than making localised extracts from the main map — note the extreme graticule curvature that occurs for insets 1, 2 and 3.

Original inset layout
and frame detail (right).
There are more graticule
lines on the smaller scale
inset map than on the
main map; scale bars are
lost in the corners and
scale ratios and titles are
missing.

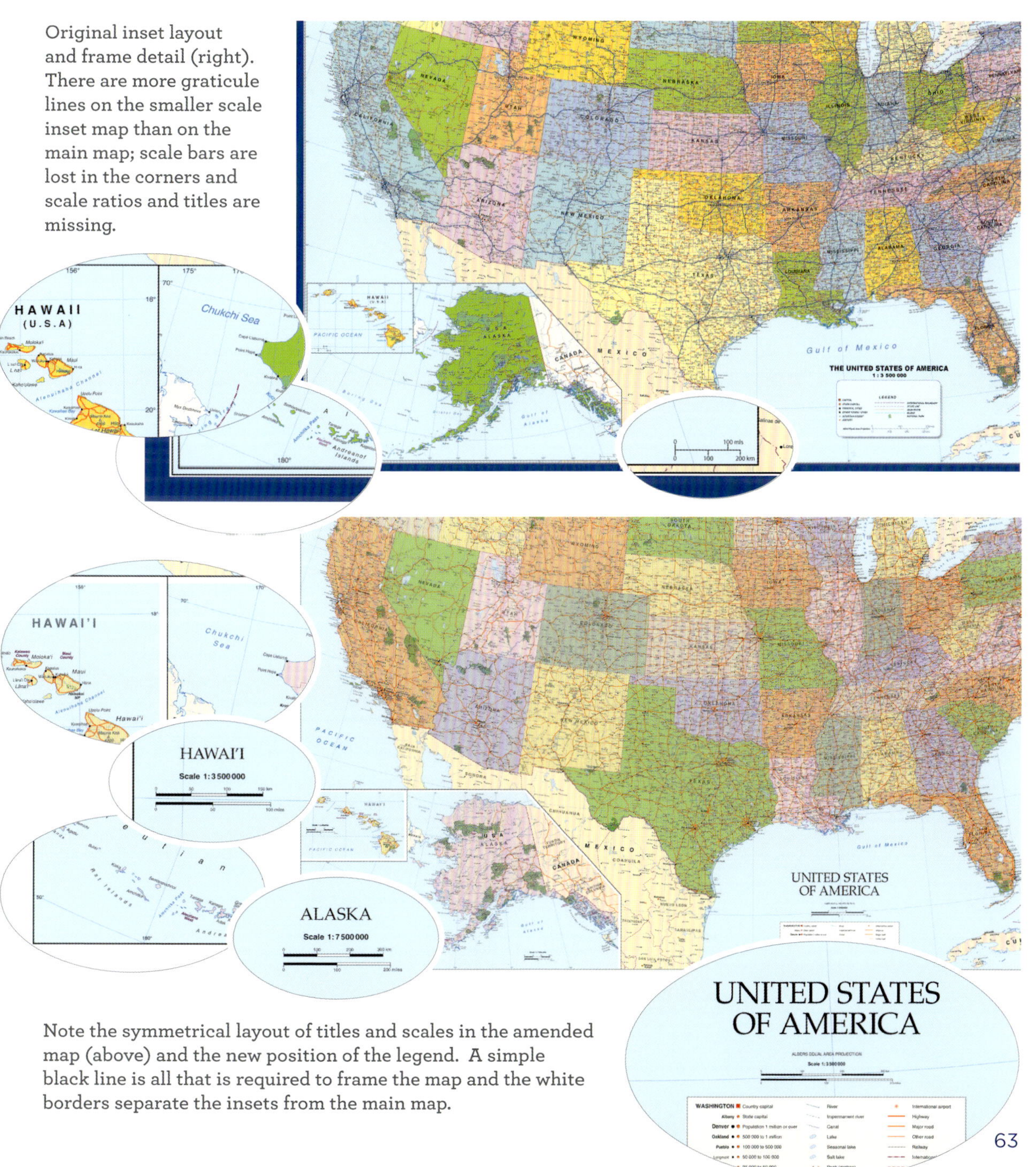

Note the symmetrical layout of titles and scales in the amended
map (above) and the new position of the legend. A simple
black line is all that is required to frame the map and the white
borders separate the insets from the main map.

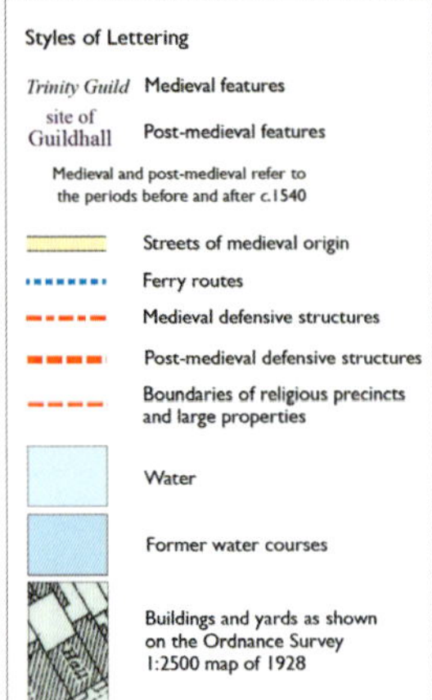



Legend for Historical Map of Hull.

© The Historic Towns Trust

**T**HE PURPOSE OF A LEGEND is to explain map symbols. The meaning of many map symbols may be obvious to the reader, especially if they are conventional or the map is part of a familiar series, but the more technical or specialist the map, the more it is likely to need a legend.

If you have a legend, it should be comprehensive, although symbols that are self-explanatory (such as coastlines) can be excluded. Basic information such as roads and rivers can also be omitted if you really do not have room. Otherwise, include all symbols shown on the map, even if familiar. Symbols should be shown in the legend exactly the same size, colour and manner as on the map.

## Legend layout

● Group related features together — for example all features to do with railways (track symbols, stations, level crossings etc.) should appear together.

● Where appropriate, group together symbols of the same graphic type. Show together all point symbols, line symbols, and area symbols and show the symbols at the same size as they appear on the map.

● Show samples of text, with the same fonts and colours as used on the map, again grouped by feature type.



Index and explanation 1:625,000 bedrock geology map.
British Geological Survey ©NERC



| ERA | PERIOD | EPOCH/AGE | | | REGIONAL VARIATION |
|---|---|---|---|---|---|
| **Million Years** | NEOGENE | N | Pliocene to Pleistocene | N2 | Red Crag, Norwich Crag and Wroxham Crag  shelly sand, clay and gravel |
| | | | Miocene to Pliocene | N1 | Coralline Crag, St Erth, St Agnes, Lenham and Brassington fms, Crousa Gravels  gravel, sand, silt and clay |
| 23 | | | | | |
| CENOZOIC | PALAEOGENE | G | Oligocene | G5 | Solent Group  silt, sand and shelly clay |
| | | | Eocene | G4 | Bracklesham and Barton gps, Bagshot Formation  sand, silt and clay |
| | | | | G3 | Thames Gp (Harwich and London Clay fms)  silty clay and sandy clay |
| | | | | G2 | Lambeth Gp (Upnor, Woolwich and Reading fms)  sand and clay, pebbly, shelly |
| | | | Paleocene | G1 | Thanet Sand Formation  fine sand |
| 65 | | | | | |
| | CRETACEOUS | K | | K6 | White Chalk Subgroup  chalk with flints   } Chalk Group |
| | | | | K5 | Grey Chalk Subgroup and Hunstanton Fm locally  clayey chalk |
| | | | | K4 | Gault and Upper Greensand formations  mudstone and sandstone |
| | | | | K3 | Lower Greensand Group and Woburn Sand  sandstone |
| | | | | K2 | Wealden Group  mudstone with limestone |
| | | | | K2 | siltstone and sandstone |
| 145 | | | | K1 | Purbeck Group  limestone, mudstone and evaporites |
| MESOZOIC | JURASSIC | J | Late Jurassic | J7 | Portland Group  limestone, dolomitic sandstone, some mudstone |
| | | | | J6 | West Walton, Ampthill Clay and Kimmeridge Clay fms  mudstone and muddy limestone |
| | | | | J5 | Corallian Group  limestone, sandstone, siltstone and mudstone |
| | | | | J4 | Kellaways and Oxford Clay fms  mudstone, locally sandy |
| | | | Mid Jurassic | J3 | Great Oolite Group  limestone, some mudstone and sandstone |
| | | | | J2 | Inferior Oolite Group  limestone with sandstone and mudstone |

## 1:250 000 Scale Colour Raster map legend

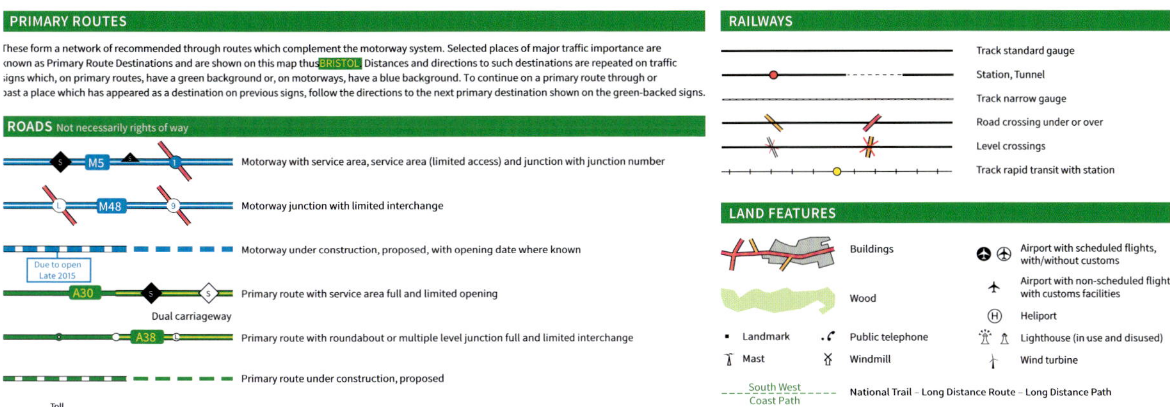



# Legends for data maps

Legends for qualitative maps need to show examples of all categories of information mapped and give all appropriate information about the statistics. Legends for quantitative maps need to indicate how the size or colour of the symbol relates to the size of the subject being mapped. More detail is given in the sections on different types of thematic map, especially chapter 34.

- State the units being used: 'percentage', 'density per $km^2$', 'number of residents within one mile of a library'.
- Units of time are often wrongly omitted, and should be stated: 'Income per person per year'.
- If the map shows change across time, you need to state dates: 'percentage change, 1991–2011'.
- Be consistent across a series of maps. Use the same abbreviations for units ($km^2$ or sq km for example) and the same pattern of legend layout.

# Scale bars

Scale bars need not necessarily be placed within the legend area. It may be more appropriate to position a scale bar with the title or elsewhere on the map. Whatever the map, a scale bar should use logical units and divisions based on whole numbers which are easily understood (so 0, 5, 10 not 0, 2.3, 4.6, etc). The complexity of the scale bar should be matched to the type of map you are producing; a very detailed scale bar for a basic location map is probably not necessary.

## OS OpenData legend



UK A1: The scale bar sits with the title in the top left corner of the map.

© Tiger Moon

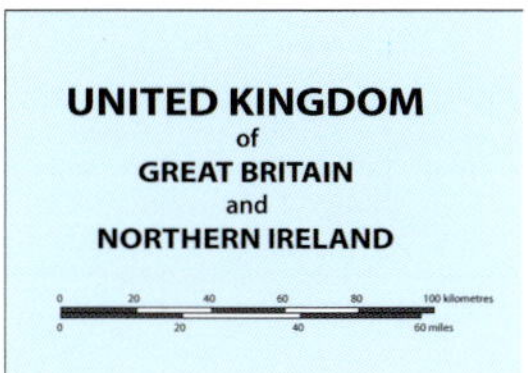

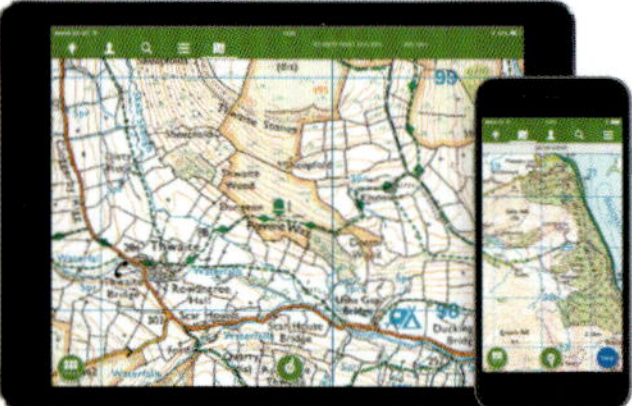

OS maps for route finding and planning of outdoor activities in a mobile app designed for use on any device.

Courtesy: ViewRanger

THE WORLD WIDE WEB, tablets and smartphones have hugely changed the way that people interact with maps. The exposure of people to printed maps is reducing as the availability of digital maps is growing. And the perception of what a map looks like is changing — from one based on the symbology of printed topographic maps, road atlases and mini atlases in diaries (which have to be paid for) to the symbology of maps delivered online (which are mostly free).

The advantages of delivering maps on screen instead of in print are considerable:

- They are cheaper to deliver.
- Maps made for print can be adapted for digital delivery.
- It is easier for the cartographer to keep them up to date.
- The map reader can look at an evolving map; a printed map is static in its information.
- Delivery of useful updates is much quicker — for instance corrections to maritime charts.
- Digital mapping can integrate and display data from different sources.
- Digital map bases are seamless, avoiding map-sheet boundaries.
- They can be interactive, and web maps can be hyperlinked.
- They offer zoom and pan functions.
- The symbology, colours, layers of information and other display parameters can be customised by the map reader, for example, to take into account colour vision deficiency or night-time map viewing.

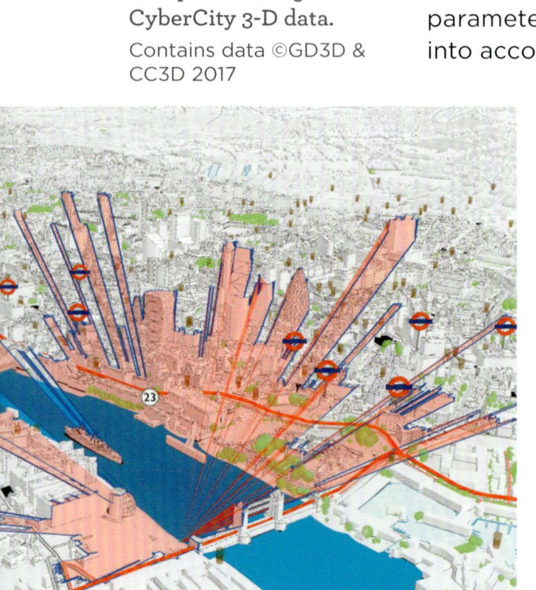

Skyline analysis of London from selected viewpoints using CyberCity 3-D data.

Contains data ©GD3D & CC3D 2017

Digital mapping and digital interaction with maps has led to categories of map and map use which never existed before.

- Collaborative maps — for example, OpenStreetMap, where volunteers contribute as well as use mapped information.
- Online analytical mapping, using the analytical functions of a GIS but with the GIS software operating remotely.
- Real-time mapping of constantly changing geographical phenomena — e.g. showing the positions of ships and aircraft.
- Static maps — the screen equivalents of paper maps delivered over the internet.

There are also disadvantages to digital maps:

- Paper maps are usually bigger, meaning geographical information can be seen in a wider context, making it easier to understand. Small screens make the map-viewing process very different and often harder.

- Paper maps have a legend to hand for easy reference.
- Different datasets are required for different scales of map because of the need to generalise map detail according to the scale of map.
- Although digital maps can integrate data from different sources, there are often issues of data compatibility: differing map projections, different scales of capture of original map data, varying dates of information, variations in reliability of data, and mismatches in data classification methods.
- The quality of mapped output is often cartographically poor, especially where the default GIS settings are used.
- Receiving maps wirelessly depends on there being a good signal; this is critical if you need frequent updates, for example to show your current position.

Perhaps the two biggest differences between paper maps and digital maps are firstly the resolution of information displayed, and secondly the viewing environment. Printed maps can show a much greater density of information per unit area (per square centimetre of the map, for example) because the separation between map symbols printed on paper can be very small and still be seen.

Computer screens and hand-held devices have a lower resolution and symbol separation needs to be larger; so computer maps need to have fewer symbols and less information. The viewing environment is also often radically different. Although looking at a map on a large computer screen can be comparable to looking at its printed equivalent, viewing it on the screen of a hand-held device allows only a small section to be seen in detail at one time. The pan and zoom functions are essential if the whole map is to be seen and understood, and the wider geographical context of the map will be much harder to grasp.

Less complex point symbols (right) for use on web mapping.
Courtesy: Steer Davies Gleave

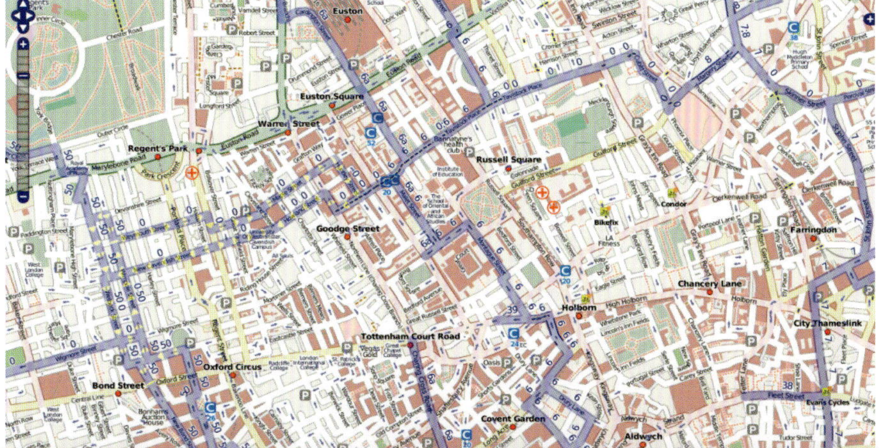

OpenCycleMap: Interactive route planning based on OpenStreetMap data.

## Design considerations for digital maps

### PDF maps

- Maps designed for printing can be made available as PDFs for screen viewing. But screen colours are mixed using RGB, not CMYK, so colours designed for printing won't look exactly the same when displayed on screen.
- PDFs printed in black and white will look different from their colour equivalents (insufficient contrast between tones is usually the problem).
- Use a scale bar instead of a scale statement to ensure that it remains meaningful when printed at a different size.

### Maps for websites

- Design maps for screen aspect ratio, to avoid scrolling.
- Resolution should be 72 to 100 dpi for web output. Smaller files will load more quickly.
- Increase the size of symbols for screen maps, and generalise more.
- Increase the weight of lines used to ensure that they are still visible.

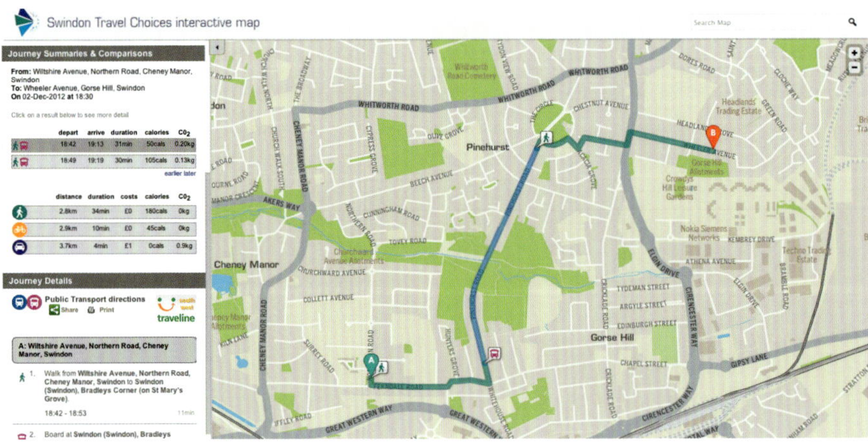

Swindon Journey Planner: Interactive route planning from Swindon Travel Choices.

Courtesy: Steer Davies Gleave

### Colour

- Colours for use on the web are specified either as RGB values (ranging from 0 to 255) or as Hex (short for hexadecimal) colours. Every colour given an RGB value has a Hex equivalent. For example, turquoise is 64,224,208 in RGB and 40E0D0 in Hex.
- Not all colours that can be seen on a computer screen can be represented by their CMYK equivalents.

> **did you know...?**
> Geospatial PDF files can hold the coordinate system and attribute information of spatial data. You can query and analyse the map from within the PDF.

> **useful tip**
> Avoid large areas of stark white on screen — it makes for difficult viewing.

- The range of colours theoretically visible on a modern computer screen is huge — more than 16.5 million. In contrast there are fewer theoretical colours available by printing CMYK: 1 million. Moreover, the range of colours that coincide between the two systems (the gamut) is limited. Any RGB colour that cannot be reproduced in CMYK is said to be 'out of gamut'.
- Computer and hand-held device screens these days support millions of colours, so you do not need to stick to the 216 'web-safe' colours. It is more important to ensure good contrast on screen maps as the subtle differences between printed colours may not be visible on screen.
- Work with fewer colours than you would for printed maps. Use fewer, well-contrasting colours.
- Maps displayed on mobiles and tablets will often be viewed in bright light, so the contrast has to be greater for symbols to be seen.

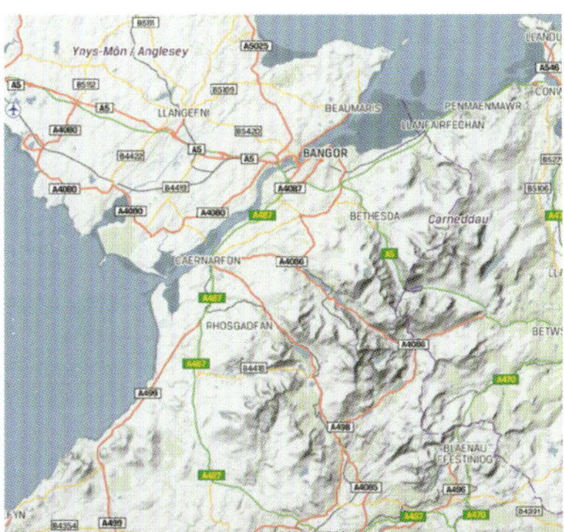

Illustration of the CIE 1931 colour space showing the different colour gamuts.

## Fonts

- When choosing a typeface, stick to common and simple serif and sans serif fonts. Nine out of ten of the most common fonts used on web pages are sans serif. Serif fonts are easily visible on screen when large, but harder to read when small, so avoid small serif labels.
- Don't use script or display fonts.
- Avoid italics and text effects such as drop shadows.
- Increase the size of the text in comparison with printed maps.

## Other considerations

- With zoomable maps, as the viewer zooms in or out, the map dataset displayed changes with the change of scale. Although the amount of generalisation between datasets necessarily changes, the same styling should be applied to different datasets so that they appear consistent.
- The resolution of screens is improving all the time, so the issue of differences between printable map information and screen-viewed information is likely to lessen.

viaEuropa: OS Open Data (below left), OS VectorMap Local (centre) and OS MasterMap Topo (below) at different scales, designed with the same 'look and feel'.

© Europa Technologies

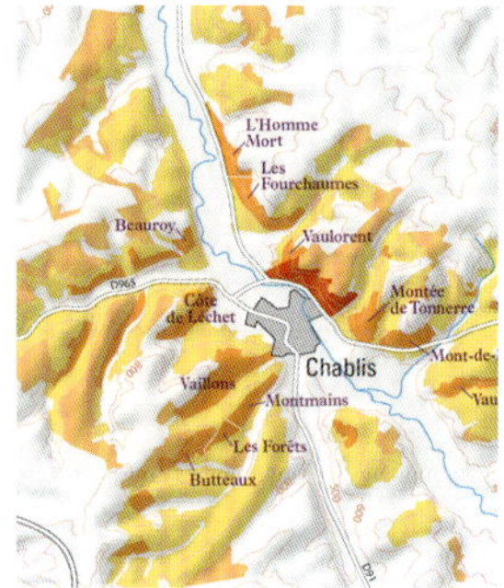

Wine map of Chablis with a shaded relief background.

© Cosmographics

OPPOSITE:
**Lampeter Geological map with the base detail in grey.**

British Geological Survey © NERC and Ordnance Survey topography © Crown copyright

**T**OPOGRAPHIC MAPS are general-purpose reference maps where all information shown is of roughly equal importance. Thematic maps in contrast show one or more categories of geographical information, deliberately making the theme more important. They include the large group of data and statistical maps, as well as specialist reference maps such as geological maps.

Thematic maps consist of two basic elements: a base map in the background to provide context, and the themed information which overlies the base map.

## Selecting a base map

Base maps provide a background reference for the thematic information. They are not the focus of the map but provide context for the theme. In some cases, the content of the base map will be obvious, for example a choropleth map must show the boundaries between the geographical units for which you are showing statistics (counties, census districts, etc). In other cases, you need to decide on all the elements to include.

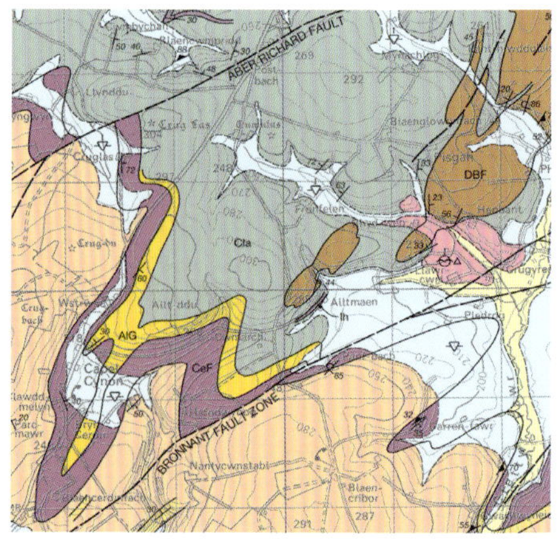

If mapping the world or a continent, the choice of map projection is critical, and almost always an equal area projection should be used when displaying statistical data. It is helpful to show the graticule on world or regional maps. For small areas, the projection is not critical.

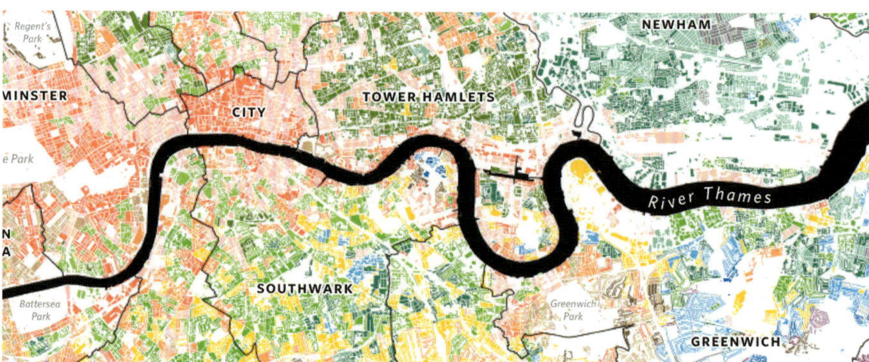


'Similarities' from London: the Information Capital.

© James Cheshire and Oliver Uberti

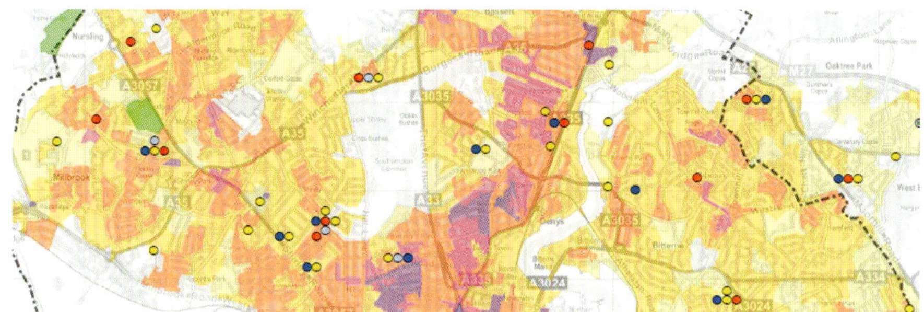

South West England:
Girlguiding GIS mapping.
Courtesy: LovellJohns

In determining the contents of the base map, the key questions to ask are:
- Does the information assist in orientating the map reader?
- Is the information helpful or essential to the map's purpose?



OPPOSITE:
Middle East: Oil
Production.
Courtesy: Global Mapping

Thematic maps of more familiar areas may require less base-map information, but you should be consistent in the selection of features shown. The absolute minimum requirements are the major land/water divisions, and major socio-political boundaries, like international, regional or county boundaries. Other topographic features to consider including are physical features such as mountains and hills, rivers, roads, and towns and cities. Thematic maps work best when the background contains enough information for context but not enough to create clutter or visual confusion. Background maps for technical maps (such as geological maps) are likely to include far more information to provide specific context for the thematic data.

If mapping an inland district, it's essential not to give the impression that the area mapped is an island. This can be done by extending features like roads and rivers across the boundary to give the idea of continuity. Detailed base-map backgrounds like OS map data can be taken over boundaries and faded out. An inset map can be used to show the mapped district in its wider context.

Base maps can be more generalised compared to their topographic equivalents. If the shape of an area is familiar, then map readers will correctly identify it.

Bedmap 2: Ice thickness.
© British Antarctic Survey

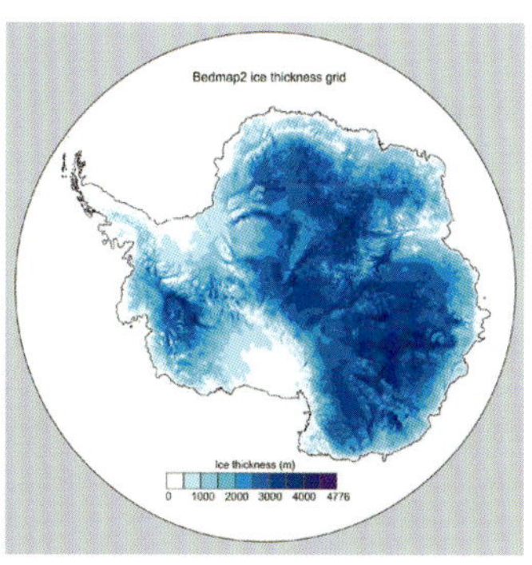

IT HAS BEEN ESTIMATED that in the region of 90% of all information has a spatial component to it — i.e. *where* something is is a component of the data. If place is an element of information, mapping may well be a good way of representing it. In comparison with a table of information, a map can show the geographical component of a set of statistics and reveal patterns that raw data alone do not. However, the way that statistics are collected, analysed and then shown on maps is not a fixed process, and the variables involved in data collection and data interpretation can result in maps which suggest wildly different patterns.

### The modifiable areal unit problem

Statistics are gathered about individuals, but conclusions are drawn about groups of people. The process of grouping individual data is called aggregation and is necessary for anonymity and to reveal spatial patterns. Census data are collected about individuals but aggregated into census enumeration districts (CEDs) — usually about 15 households. CEDs are further aggregated into larger units such as postcode sectors, electoral wards or divisions, local authorities. When census results are published, the general characteristics of the CED or area can be described: average income, level of education, average age, and so on.

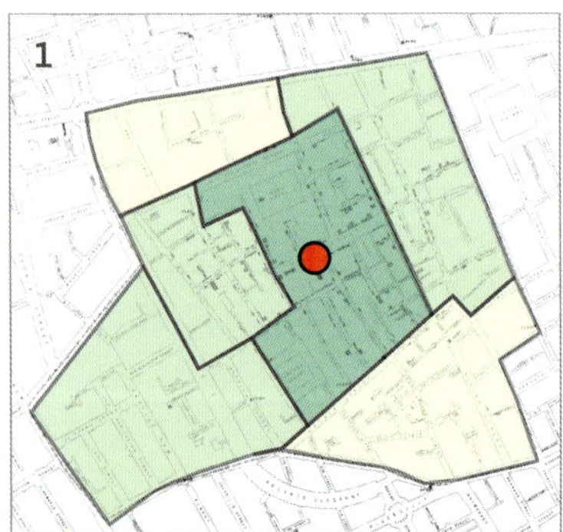
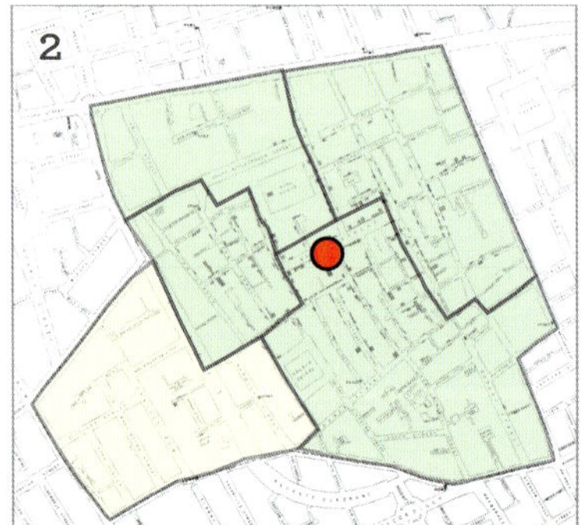
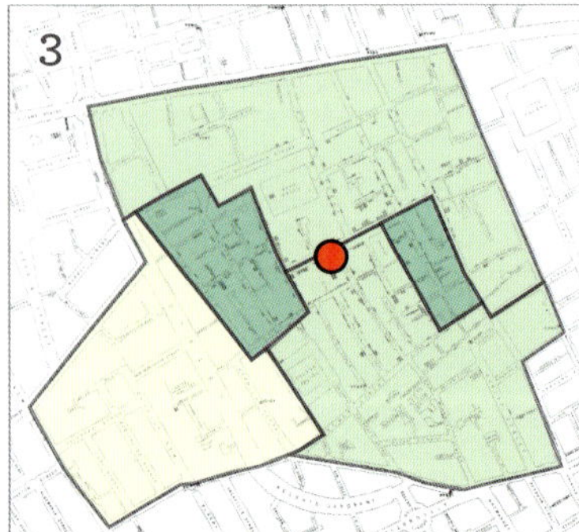

John Snow was a physician working in Soho, London during the 1854 outbreak of cholera. He hypothesised that cholera was a water-borne disease and that the outbreak was related to a polluted water pump. He plotted the location of water pumps and incidence of cholera on a base map; victims were shown by a small black mark. A visual examination of the map showed a cluster around the pump in Broad Street, and he arranged for the pump handle to be removed to stop people using it.  The incidence of cholera immediately declined.  But if Snow had produced a choropleth map from his raw data, he might not have drawn the same conclusion, depending on where he put the boundaries of the enumeration districts. As the three maps above show, different conclusions can be drawn from the same data.  Map 1 shows a clear cluster of cholera around the Broad Street pump; Map 2 is inconclusive; Map 3 firmly suggests a negative correlation between cholera and the Broad Street pump. All three maps are legitimate interpretations of the data. (Example after the work of Mark Monmonier.)

However, the way in which data are aggregated can radically affect the conclusions that are drawn about the population. One CED may appear to have average income, but it may in reality have twelve poor households on one road, and three rich ones in the next, joined together into a single arbitrary unit.

The boundaries of the areas for gathering geographical data are almost always arbitrary, even if logical, and as a result the conclusions you can draw about the inhabitants are influenced by where the boundaries are drawn. This problem is called the modifiable areal unit problem (MAUP). Conversely, the more that data are aggregated, the less likely it is that any one individual within the aggregated area exactly fits the description of 'average' for the aggregated area; this tendency is known as the ecological fallacy.

## Sampling bias

Problems can also arise through the way that information is sampled. It's rare to find out information about every individual in an area, and normally a statistical sample of the total population is taken. As electoral opinion polls have recently shown, well-run and conscientious polls can get it wrong. In spatial statistics, there are many different methods of sampling. Problems can arise by using a sampling grid that is too coarse which misses fine detail; having too small a sample size; using a regular grid that gives a false impression of the sampled area; and random sampling may, through bad luck, give a false impression by hitting on subjects that just aren't representative of the total population.

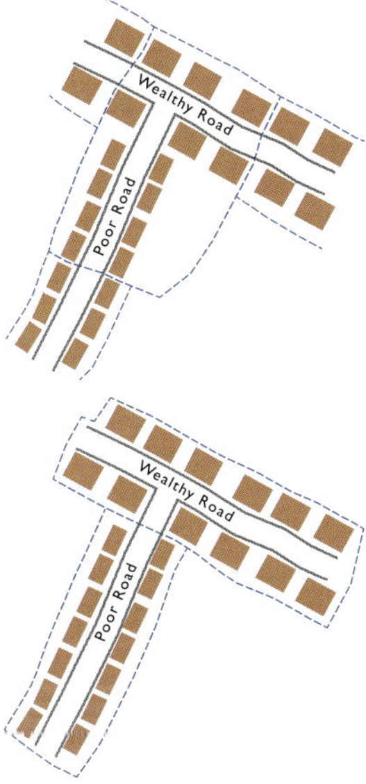


The modifiable areal unit problem: two different conclusions drawn from the same data by varying the boundary of the sampling area.

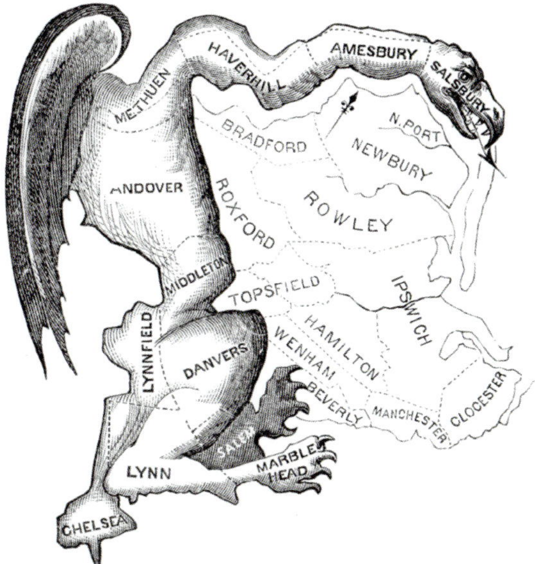


Perhaps the most famous example of the deliberate manipulation of boundaries is the creation in 1812 of an odd-shaped electoral district in Massachusetts, USA. The state's governor, Elbridge Gerry signed into law a bill to redraw the electoral boundaries of the state, designed to favour the Democratic-Republican party which he represented. One of the districts in the Boston area, Essex South, joined together several smaller counties to form a district which was likely to contain voters favourable to his party. The district's shape was contorted, and was described as being like a salamander. The Boston Gazette newspaper coined the term 'gerry-mander' as a combination of the governor's name and the word salamander for the shape of the district. Gerrymandering has entered the language to describe any deliberate attempt to distort electoral results by applying the modifiable areal unit problem.

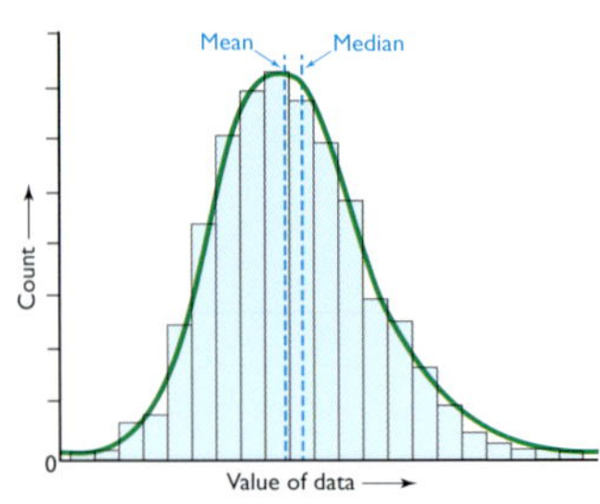

Normal data distribution

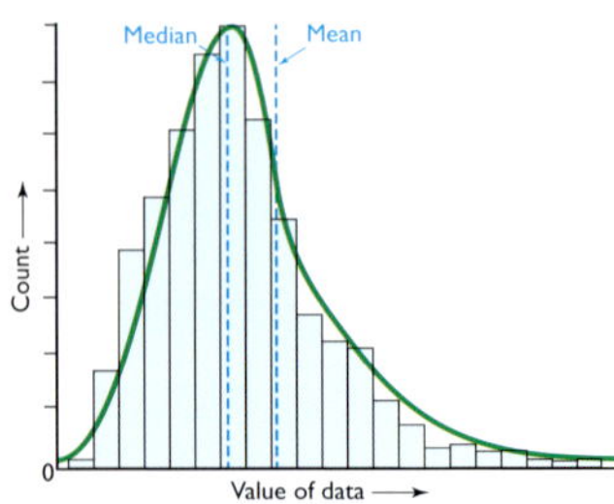

Positively skewed data

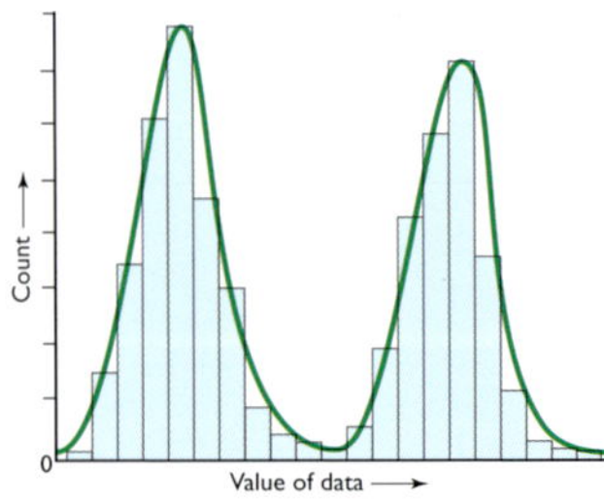

Bimodal data distribution

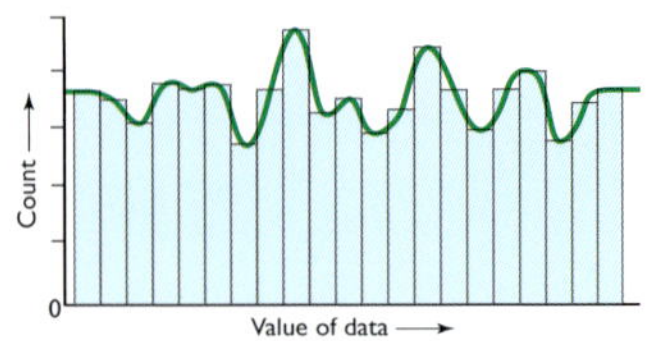

Random data distribution

CARTOGRAPHERS ARE OFTEN REQUIRED to map a table of data. If there is too much data to map each item, you need to engage in data classification. Classification is the process of dividing data into groups which have meaning according to the statistics that are being mapped, and meaning to the map's audience. The purpose of mapping a set of statistics is to visualise the geographical patterns that the statistics reveal, and good classification can be an aid to that process.

There is no single, foolproof way of classifying data, and therein lies the potential to mislead an audience. Different methods of data classification can result in very different maps, all derived from the same dataset. You therefore need to understand something about the different methods of classification. It is critical that the characteristics and significant aspects of the mapped phenomenon are well represented.

### Dealing with a set of statistics

Examine them; get a feel for them; ask what they show. Determine the highest and lowest values and therefore the data range. If you have some skills with statistics, work out if the data are *normally* distributed, or *skewed* (heavily biased towards low or high values), or *bimodal* or *diverging* (such as population growth and decline in a region).

### Methods of classification

Although there are many ways to classify data, there are four broad groups of data classification used by cartographers.

- Arbitrary. Class divisions are at regular, rounded numbers (100, 200, 300, 400, etc.). This is an easy method to use, and the legend is understood by the map reader, but it often reveals very little about the data.
- Exogenous. Class divisions are imposed from outside. For example, income levels may require mapping to agreed standards where data divisions are set nationally. The 'poverty line' may be a change point, as may be the threshold for higher-rate taxpayers for a map of income by area.
- Idiographic. Graphing the data may show 'natural' break points and clusters in the data. This method is often used when mathematical methods fail to produce a useful classification. The intervals then chosen usually look very odd and are not well understood by the map's audience.
- Mathematical or serial methods. These are the most rigorous methods.

### Mathematical methods

Mathematical methods include: equal intervals; intervals based on medians, quantiles and standard deviations as determined by statistical analysis; and intervals based on arithmetic and geometric series.

- Equal intervals. Often the default output of GIS analysis and easily understood by map readers although rarely showing much about the data. Equal interval classification often results in one or more empty data classes but can be useful if the spread of stats is even across the range.
- Medians and quantiles. Data are set out in rank order, and the median value found. The median can be a break point for two data classes. Otherwise the ranked order can be divided into 3 (terciles), 4 (quartiles), or 5 (quintiles) etc. groups with equal numbers of observations in each class. Quartiles reveal the inter-quartile range (the 50% of observations around the median). Revealing for stats with an even spread across the range, but often the interest lies in the extreme values (e.g. high or low mortality).
- Means and standard deviations. If the mean is close to the median and there are very few extreme values, your data *may* be normally distributed. However, most geographical data are *not* normally distributed. The mean is calculated and may be a break point between classes. Standard deviations can be used to determine class intervals; fractions of a standard deviation can also be used. An example of data that might be normally distributed would be journey-to-work travel-time.
- Arithmetic and geometric series. Appropriate if your data are slightly or strongly skewed. Arithmetic series are useful when the data are slightly positively skewed and you want smaller intervals at the lower end of the data. They are based on class intervals that get bigger by a set amount that doubles, then triples, quadruples etc. Geometric series are suitable for very strong positive skews in data; data classes are determined by powers of a number that form the classes — for example $10$, $10^2$, $10^3$, $10^4$, etc. giving class break points of 10, 100, 1000, 10,000 etc.

Data values that are a long way from the 'body' of stats can be hard to deal with, as they have to be accommodated but can skew a classification. If you exclude them from the classification, you can show them as symbols or figures to the map with explanations in the legend: e.g. 'Israel, Jordan, and Lebanon produce less than 500 barrels per day'.

## Number of classes

You should have no fewer than 4 classes (too generalised) and probably no more than 8 (you can't tell the difference between symbols). For choropleth maps, it's hard to distinguish more than 6 different colours or tones; more than 5 or 6 differently sized proportional symbols may be difficult to identify. Bimodal series, where you are using two different colour ranges can accommodate up to 4 or at most 5 classes each side of the neutral break point.

Europe Density of Population: A data range of 3 to >18,000 people/km² creates data classification problems. With equal intervals, 53 of 55 countries are in Class 1, and 53 are within one standard deviation of the mean. Quantiles show a far more revealing pattern.

Equal intervals

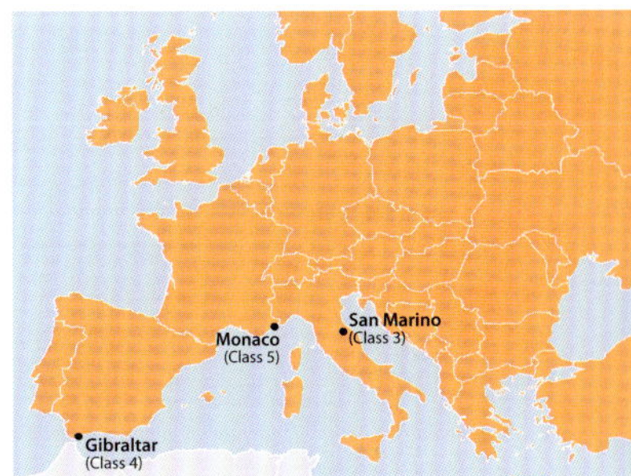


Standard deviation

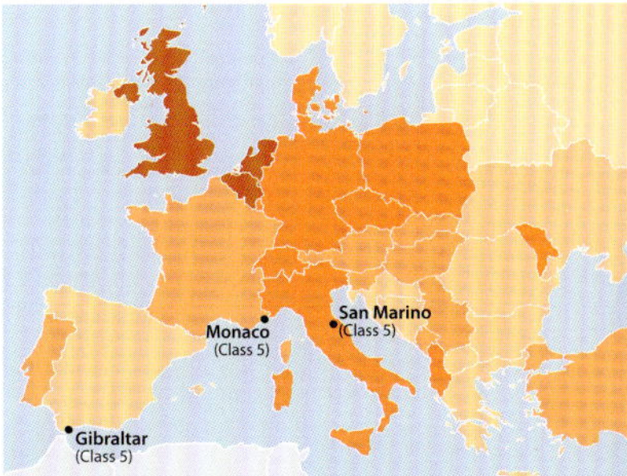


Quantiles

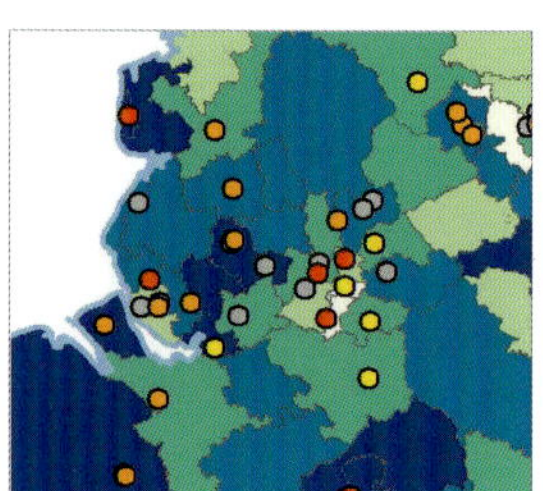

Choropleth maps can be combined with point data. Here, Obstructive Sleep Apnoea (OSA) risk factors are shown as a choropleth using health areas as the basis. Shading represents the expected prevalence of OSA within each area (as a percentage of the areas' population with the condition); point data represents sleep clinics, also classified.

Courtesy: LovellJohns

Environmental World: Annual Change in Forest Area, 2000-2010.

© Global Mapping

THE CHOROPLETH MAP is a 'value by area' map. It is a common way of mapping data gathered by area, and can be a very effective way of conveying the spatial variation in a set of statistics. The choropleth map combines a base map, showing the areal units on which the information was gathered, with infill patterns or colours which represent the data class assigned to the area.

## Data suitable for choropleth mapping

Choropleth maps show data gathered by predefined area, such as census enumeration districts, local authorities, wards, counties, or countries. Data are often aggregated to give information about larger spatial units. Ideally, data have to be available for each area that is to be mapped.

Choropleth maps should only be used to map *ratio* data (population per square mile, income per capita, etc), and never *absolute* data. For example, mapping the number of births in an area is largely unrevealing because the number is related to how many people live in a place. Far more informative is the birth *rate* — the number of births per 100,000 population per annum.

## Choropleth map compilation

The sequence involves creating a base map, classifying the data and assigning colours or infill patterns to the areas. The base map will show the boundaries of the units on which the data have been compiled, as well as sufficient topographic information to give it context. The data are then classified. Choropleth maps should ideally have no more than 6 classes of data, perhaps 7 at the most, as it is very hard to distinguish more categories on the map; even if they are clear in the legend, it may be hard to distinguish them on the map.

Density of the mapped phenomenon is shown by varying colour or pattern. The basic principle of the shading is that, as data figures go from least to most, the colour sequence goes from lightest to darkest, or least to most saturated — lighter colours represent lower data values and darker colours represent higher values.

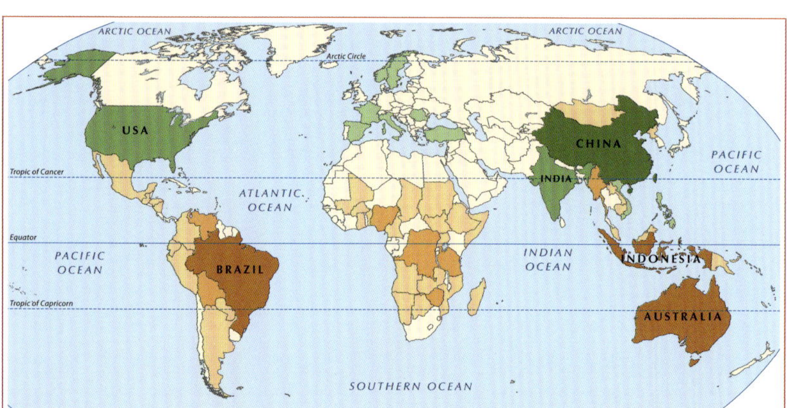

There are a number of possibilities for colour sequences, but the two most common solutions are using a single hue (colour) and increasing its darkness, or using a narrow range of colours and progressing through them — for instance from blue through green to yellow, or blue to purple.

For bimodal or diverging series, you can use two hues and vary their darkness, or two complementary colour sequences. The lighter values should be in the middle since it represents a neutral point between the diverging sequences.

If you are mapping in black and white, varying shades of grey can be used for infill patterns, but do not use more than 70% grey or it will look like solid black. Different fill patterns can also be used, but some feeling for the patterns becoming more intense should be shown.

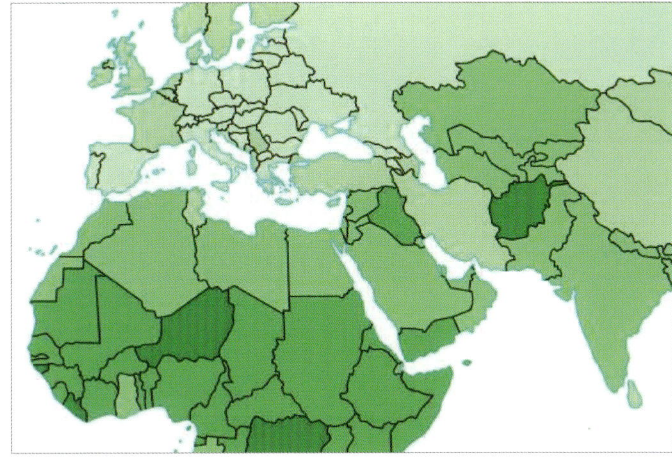
Times Comprehensive Atlas of the World: Total Fertility Rate, 2005–2010.
© CollinsBartholomew

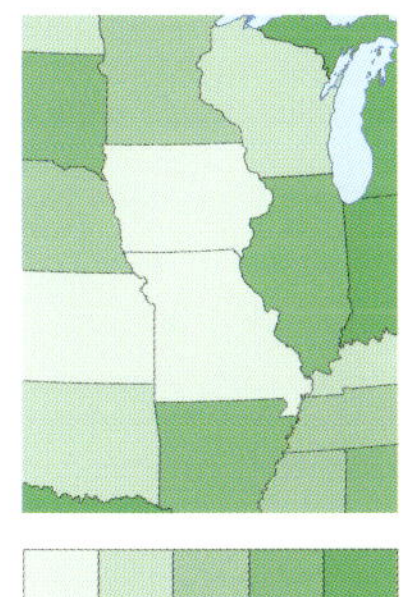
increasing saturation

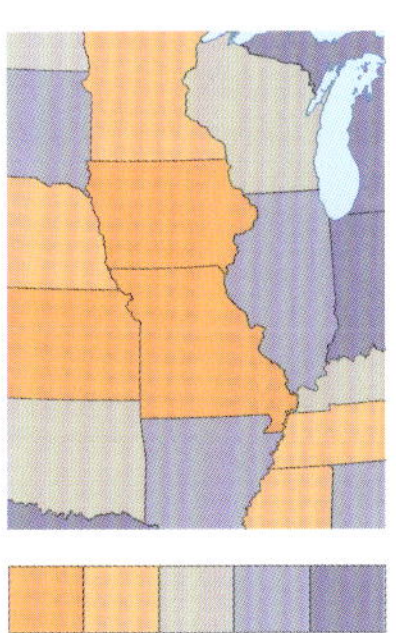
hue transition

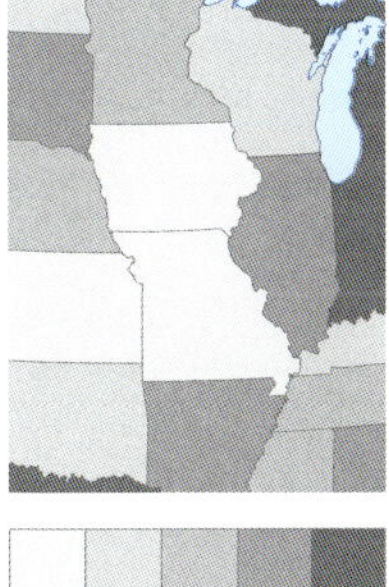
diverging colours

increasing value

## Legends for choropleth maps

A legend for a choropleth map should show all the colour fills or patterns used and should label the data range.

- Decide if the highest value will be at the top or bottom of the scale; it is usually at the top.
- Label each entry with its range (e.g. 101–200) or label the breaks between classes.
- State the units being used, especially for ratios, e.g. 'percentage', 'density per km$^2$', etc. If the map shows change across time, you need to state dates, e.g. 'percentage change, 2006–2016'.

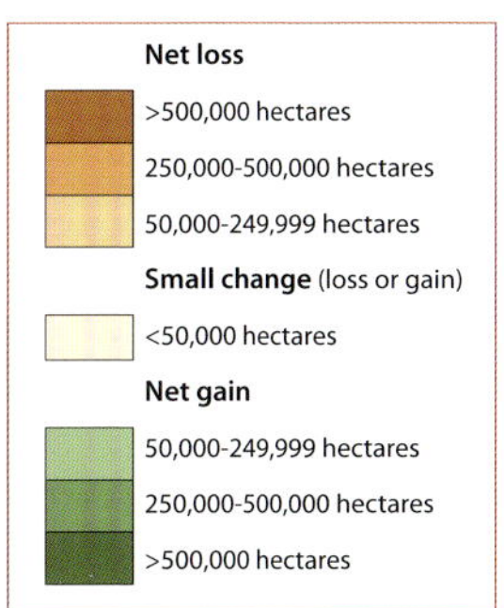

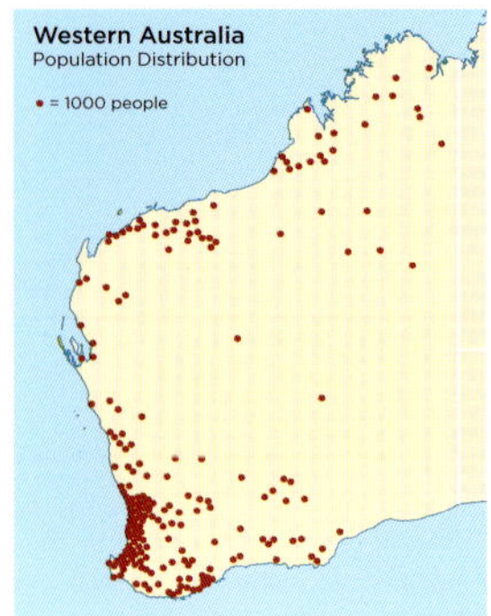



**D**OT MAPPING CAN BE an effective means of showing the relative density of a phenomenon within an area. The idea of a dot map is to give an impression of density and spread of the subject being mapped and, as statistical maps, they are easily understood by the map reader. The variables that the cartographer has to choose are the value of the dot and the size of the dot.

## Data suitable for dot mapping

Dot maps are used to show absolute data, rather than ratio data. For example, they can show human population or agricultural production. They work best when the subject being mapped has a wide distribution across the mapped area.

## The base map

The base map should show enough topographical information to give the theme context. However, it does not necessarily need to show the areal units on which the data were gathered. For example, a population dot map might show information gathered by electoral districts but it is not essential to show those on the final map.

## Choosing the dot value

Each dot represents a set number of real world objects. It can represent a single object, but usually represents a number of objects — for example 1 dot = 50 or 100 people. It is desirable to have 2 or 3 dots in the area with the lowest data value, and in the areas with the highest data values it's good if the dots look like they are just beginning to coalesce.

To calculate the dot value, examine the range of the data and work out a ratio of units per dot. Then calculate how many dots at that ratio there would be in the area with the highest value and check to see if you can accommodate that number in the area; if not review the dot value. The dot value cannot be greater than the lowest item of data otherwise no dot will appear in the area with the lowest data value.

A dot size that is too small (left) makes the distribution appear sparse; too large (centre) and it is difficult to pick out centres of population. The intermediate dot size (right) allows local centres to be identified without making the rest appear empty.

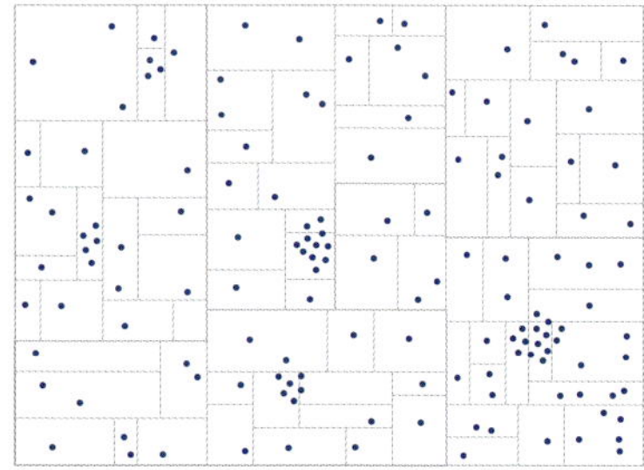
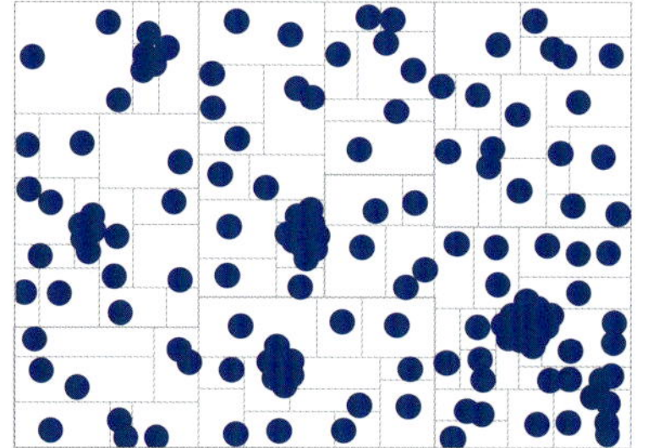
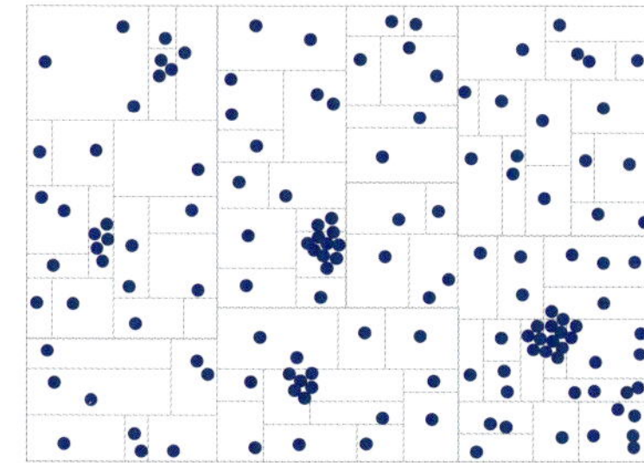

## Choosing the dot size

Finding the right-sized dot may be a process of trial and error. Dots need to be big enough to be seen (especially in areas with only a few dots), but not big enough that they overcrowd the most dense areas. The map should not look sparsely dotted.

## Dot placement

The general idea is to place the dots to correlate with the real distribution, but without giving a false impression of concentration. Use your base map of the areas which the data relate to and put the right number of dots in that area.

- Spread the data around within the area, going up to the boundary of it. Once the boundary lines are removed, it looks odd if there are gaps.
- Take into account real-world geography. If you know of a limiting factor in the real world, don't put dots there. For example, if there is a lake in area, it's reasonable not to put dots representing population in it.
- Otherwise, spread them randomly within the area; don't arrange the dots geometrically.
- Dots in the highest value areas can begin to coalesce. They can be separated by halos if you want to avoid a dot blob.

## Legends for dot maps

A legend for a dot map should have a clear statement of the dot value — for example, 1 dot = 50 people. If the map shows the boundaries of the statistical areas on it, the legend should also show at least 3 identical squares or polygons with three representative dot densities: low, medium and high. Map readers never count the number of dots; it's the overall impression of density and distribution that is seen.



Los Angeles: Race and Ethnicity, 2010: Red is White, blue is Black, green is Asian, orange is Hispanic, yellow is Other, and each dot represents 25 residents.


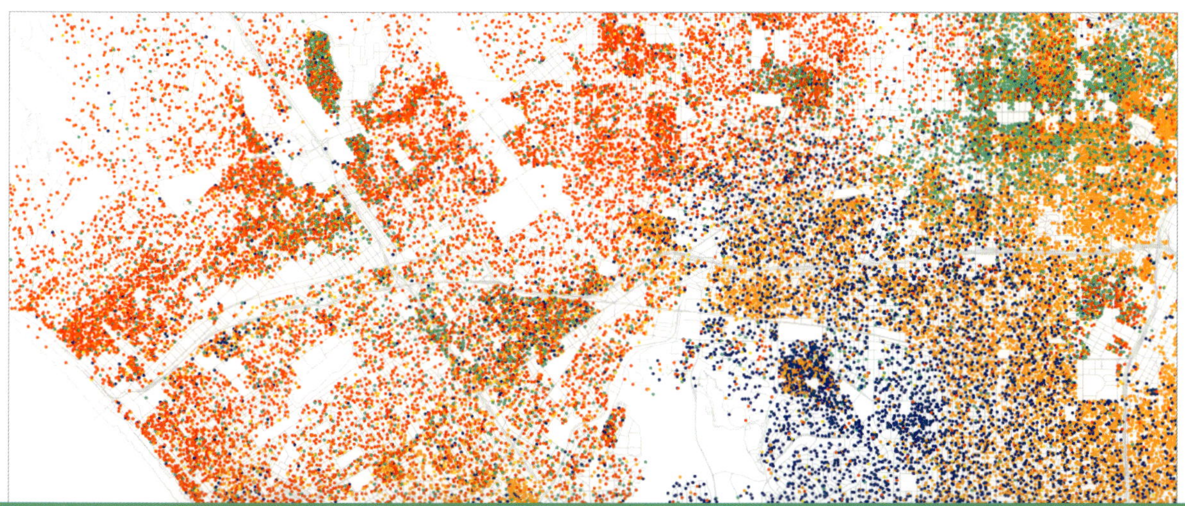

ISOLINE MAPS REPRESENT QUANTITIES by lines. The most familiar isoline map symbols are height contours — lines which connect points of equal elevation above sea level. Other examples which may be familiar are isobaths (submarine contours), isobars (lines of equal pressure) and isotherms (lines of equal average temperature). Isoline maps represent *continuous* types of data (i.e. occurring everywhere), and not *discrete* data (only to be found at points).

Isoline maps come into two groups: those that represent data that really exists at points, and those that represent data that don't exist at points. Isoline maps are often called isarithmic maps.

Data that really exist at points include heights and depths, temperatures, air pressure, sunshine and rainfall. They are all continuous phenomena and any point can be given a value. For example, lines of equal rainfall per annum (isohyets) can be drawn using data obtained from weather stations. If isohyets are shown on a map with 10mm per annum increments, it's theoretically possible to find a point between the lines that represents a 5mm per annum increase or decrease; this sort of data can be said to exist at points. Isoline maps that show real point data are called *isometric* maps.

However, a lot of data that we have are ratio data that relate to areas, and not points. For example, population density and crop yields are statistics that relate to areas: people per square kilometre, tonnage of barley harvested per hectare. Although the data relate to an area, you can still make an isoline map from them, if the areal units are sufficiently small, roughly equal-sized and shaped, and are themselves continuous. Isoline maps showing data that don't strictly exist at points are called *isoplethic* maps.

Isolines are almost always shown at equal intervals (2, 4, 6, 8, etc. and not 2, 5, 8, 12). Having small intervals allows for the depiction of subtle differences in areas of small change, but may lead to very crowded maps in areas of high change: for example, population density usually increases hugely in urban areas, leading to many lines in a small area of a map.

The quality of an isoline map depends on the frequency of sample points: the more data you have, and the closer spaced are the sampling points, the more reliable is the map. Widely spaced data points can lead to a map which implies more precision than it merits. You also need a good geographical knowledge of the area to apply common sense to the interpolation of the data.



Travel times for tsunami waves after the 2011 Sendai earthquake.
Courtesy: NOAA

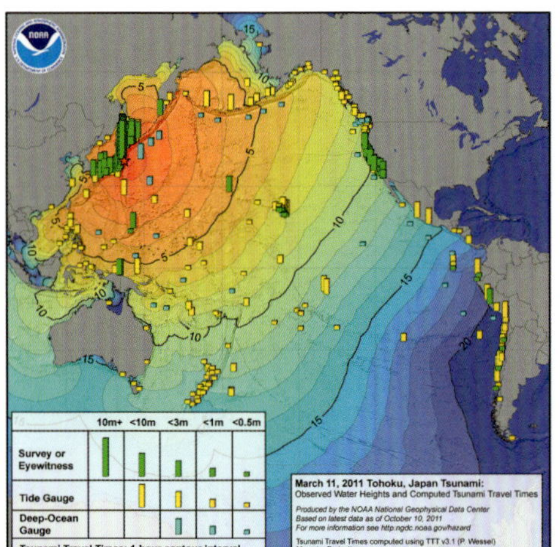

## Isometric maps

Isometric maps are used to represent data that are continuous and varying. Making an isometric map is a question of joining the dots of equal value, and interpolating between them. The interval between the lines has to be determined.

For example, a map of isotherms may have lines every 1°. Data come from weather stations and show temperatures to 0.1° precision. You will need to interpolate the position of points of whole degrees, since most items of data will be fractions of degrees. This is done by calculating where the theoretical point of a whole degree lies between two points of known value. For instance, if two points have a known value of 10.8° and 11.5°, the 11° isotherm would pass nearer to the 10.8° point than the 11.5° point. The process of interpolation is often automated.

## Isoplethic maps

In isoplethic mapping, data relate to areas: real geographical units like census enumeration districts, wards, travel-to-work areas, agricultural statistical areas, etc. or to artificial areas superimposed on the real world as a sampling grid. Hexagonal sampling grids are best. Isoplethic maps are good alternatives to choropleth maps if it's reasonable to assume that the data are continuous and smooth across an area.

A point is assigned to each area that represents the data value for the whole area, and the points are then joined up, as for an isarithmic map. The smaller the areas, the more frequent will be the data points and the closer will be the approximation to an isometric map. Again, the actual position of isopleths usually has to be interpolated. The position of the representative point within the area can affect the map; it is usually, however, placed in the centre of the area.

Isolines are familiar to most map readers, although many have problems in interpreting them. Isolines should be labelled with their values, although labelling just 'index' isolines is fine if they are spaced closely together.

Isolines can be made easier to interpret by filling the spaces between them with colours. In the case of height contours, the result is called hypsometric tinting. Isolines can be grouped together into bands of values for infilling.

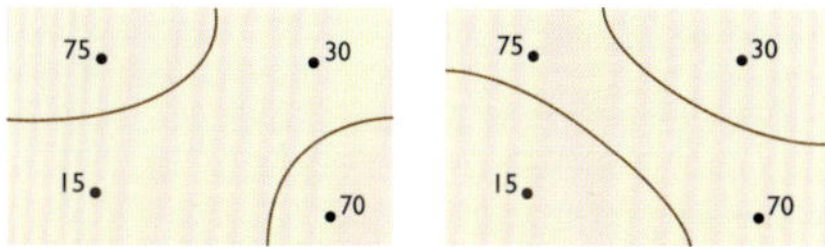


The Geologic Evolution of Saudi Arabia: Depth to Arabian Platform.

© Global Mapping

Interpolating between points of known value is not foolproof. Most methods assume that the change between two points of known value is linear (it has the same rate of change throughout), but in the real world this is often not the case. Knowing something of the geography of the area may help.

There is a specific problem where four data points are arranged in a rectangle, with two low values paired diagonally opposite two high values. When interpolating a contour between them, there are two possible routes that the line can take, both valid given the data values. Reference has to be made to other points to avoid the wrong interpolation of the line. Again, knowledge of the geography is useful.

QUALITATIVE MAPS SHOW CLASSES of information which do not have any particular hierarchy, and all classes of data are of equal importance. That does not mean that the data are not structured or grouped, but whereas quantitative data show a progression from low value to high, qualitative maps generally do not show one object as being more important than another — just different.

Qualitative maps don't involve numbers. They include land use and land cover maps; vegetation, geological and soils maps; climate zone and time zone maps; language and ethnicity of a population maps; trade routes; maritime charts; planning zone and resource maps; and many tourist maps.

### Points

Qualitative point symbols should look about the same size and weight to show different classes of information. They are often either geometric shapes, conventional or mimetic (imitating some characteristic of the feature). They can be varied in shape or colour,

Map readers will usually not need reference to a legend to understand many point symbols, because they are familiar or self-evident. However, unusual, irregularly occurring, and technical point symbols, such as those on maritime charts and geological maps, need explanation and more reference to the legend will be made.

### Lines

Qualitative maps use lines of similar characteristic (width, visual weight, and complexity) but they are varied in colour or design to show differences. Many qualitative line symbols are conventional, such as railways or waterways. Classes of data which have characteristics in common (e.g. related to water) should have related symbols, such as having the same colour.

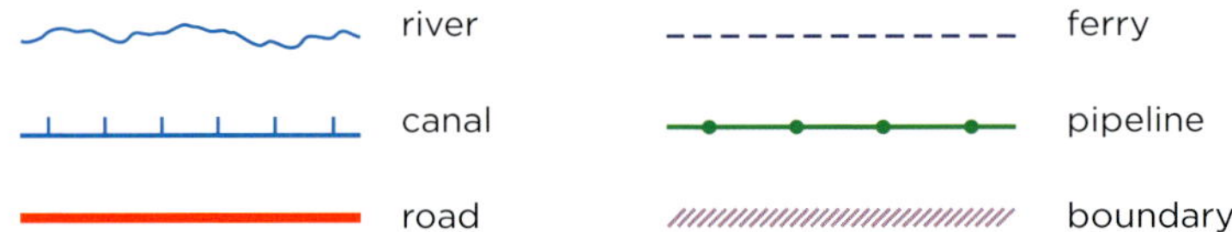


South Northants Heritage Trail: Symbols in colour-coded themes — walk information in red, heritage in brown and tourist in purple.

© Global Mapping

Charnwood Forest Regional Park: Footpaths grouped in the legend.

Courtesy: Global Mapping

Some qualitative lines need to imply a hierarchy of importance. For example, administrative boundaries need to reflect the status of the unit that they form the boundary to: electoral division, district council, and county council boundaries need to be distinguished by increasing the line weight, complexity of symbol, or the length of the peck on a pecked line.

Flowline maps indicate routes but not quantities. They include actual movements (for instance shipping or airline routes, ocean or air currents and patterns of migration), and notional movements (flow of money and ideas, for example). Either actual or indicative routes can be shown; for world and other small-scale maps, it's usually an indicative route.

## Areas

Use different colours or fill patterns which look about the same visual weight for qualitative information. On colour maps, this is done by varying the lightness and saturation of different hues (colours) to give the impression on the map of similar weight. This is quite straightforward to do on maps which have similar-sized areas to fill, but where the areas vary greatly in size, or where you have many small areas intermixed, this becomes a problem. You may need to increase the saturation (intensity) of one colour to ensure that it is visible in small areas of the map. If you have more than about five classes of information (and colours), it may be hard to tell the colour of small areas anyway. In this case, the addition of a number or letter may help. The number or letter should be added to all areas that have the same fill, and should appear in the legend.



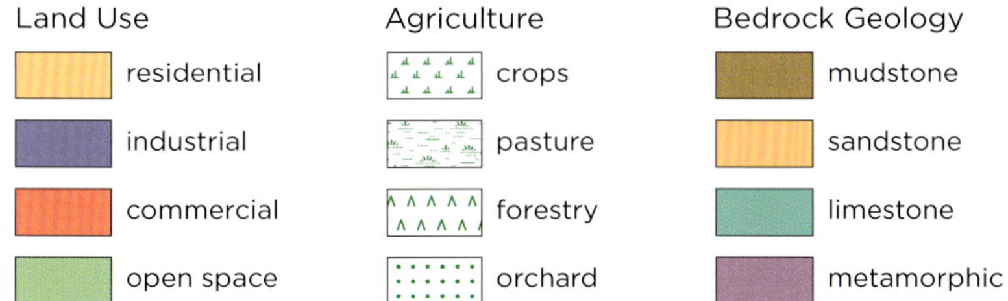

| Land Use | Agriculture | Bedrock Geology |
|---|---|---|
| residential | crops | mudstone |
| industrial | pasture | sandstone |
| commercial | forestry | limestone |
| open space | orchard | metamorphic |

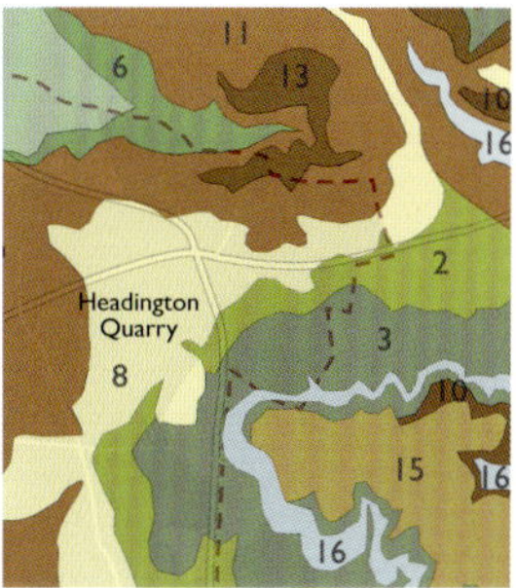

Bedrock Geology from Historic Towns Atlas, Volume VII, Oxford: Adding numbers helps to identify the bands of rock in the legend.



Classes of data with something in common can be assigned variations of one colour group. Some colours are associative (e.g. blues for water, green for forests), but often it is not possible to use associative colours for all categories, and two different categories may have associations with the same colour. Some categories of information have set symbols that should be used: for example geological maps have designated symbols for many rock types. All categories of area on the map should appear in the legend. Group similar classes of feature together.

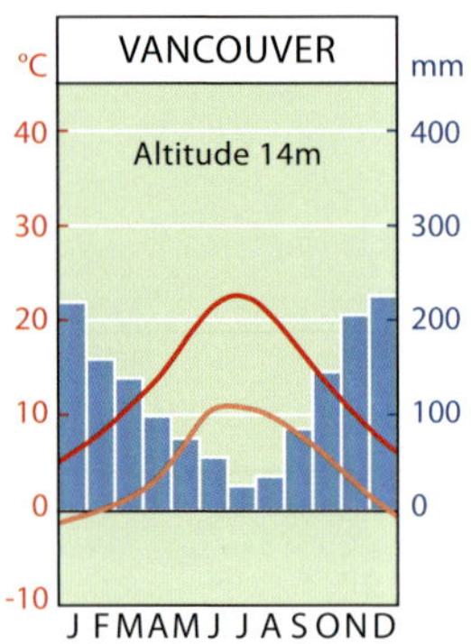





Nested circles: A neat way to show proportional symbols in a legend.

QUANTITATIVE MAPS SHOW DATA where hierarchy and quantity are significant; quantitative maps involve numbers. Maps should show differences in quantity by means of varying the symbols in size, weight and colour to indicate changes from low to high.

### Point symbols

Point symbols include proportional symbols and symbols which remain constant in size but which show numerical information — for example pie charts. Graduated and proportional symbols increase in size in proportion to the quantity being mapped. They include 1-D symbols where the height of the symbol varies (such as bar charts), 2-D symbols where the area of the symbol varies (including proportional circles, squares and other shapes) and 3-D symbols, where the apparent volume of the symbol changes (including spheres, cubes and pyramids).

### 1-D symbols (height)

1-D symbols work well where there is a fairly small range of data. To construct a bar chart, firstly work out the maximum length of a bar that can be put on the map. Find the highest value and lowest value items of data. The highest value will be represented by the longest bar. Calculate the scaling factor, and apply it to all other data items, starting with the lowest value to ensure that it is still long enough to be seen. Specific data values can be added to a bar chart if there is room.

Environmental World: Land Degradation.

© Global Mapping

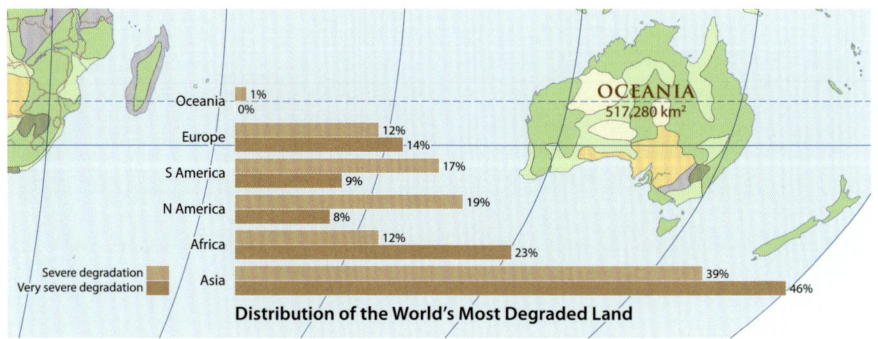



### 2-D symbols (area)

2-D symbols work well when there is a bigger data range, as the size of the symbol is related to the square root of the data value. They include proportional circles, squares and triangles, and custom-made symbols which may be imitative. Proportional circles tend to work best as they are familiar, visually stable and sit well on maps.

To construct a proportional circle, firstly work out the maximum diameter of a circle that can appear on your map. Find the highest value and lowest value items of data. The highest value will be represented by the largest circle. Calculate the square root of the highest data value, and from it determine the scaling factor. Find the square roots of all the other data values and apply the scaling factor, starting with the lowest value to ensure that it can still be seen.

Proportional squares are worked out in the same way: the square root of the data value is proportional to the length of the side of the square.

It may be easier to classify the dataset, and group into several groups (range grading) and then use a representative circle to depict all items in each group. This avoids having to construct unique circles for each data item.



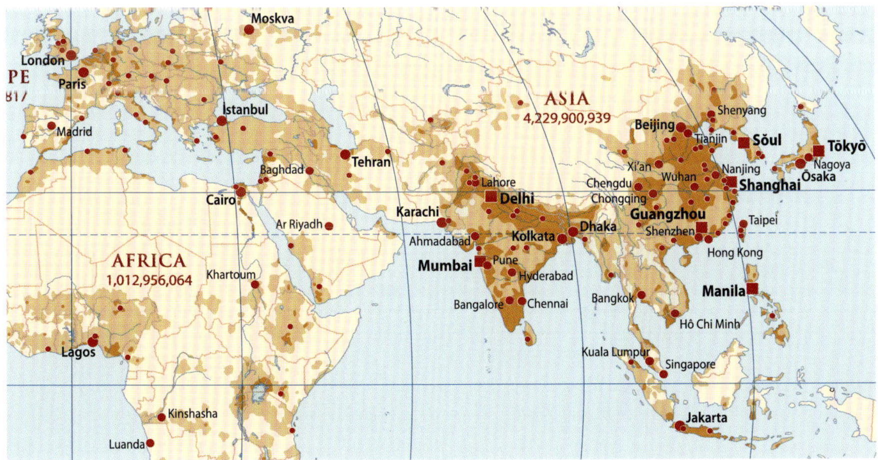


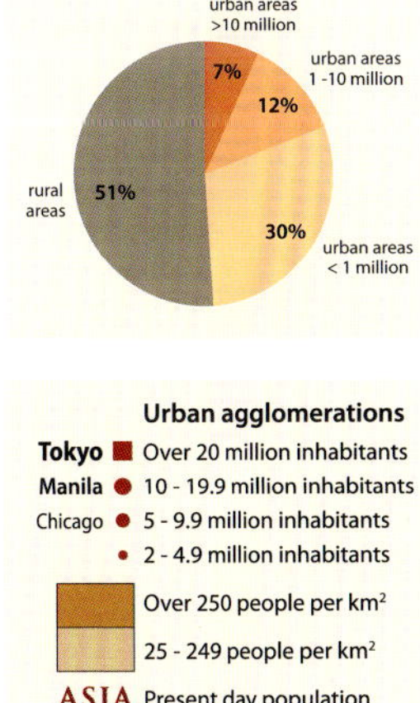
Distribution of Urban and Rural Population.


Environmental World: Urbanisation.
© Global Mapping

Proportional symbols can be subdivided (pie charts) to show the make-up of the item of data; for example different forms of agricultural production. Keep the contents in the same order in each pie chart, and start from 12 o'clock. Either label each sector on the pie or create a separate legend if there are several. The data values can be added too.

When placing circles on a map, for the most part it is fine to overlap symbols, especially when there are many of them in one area. But ideally they should all have a good proportion of the circle showing — enough to appreciate the relative size of the symbols.

OPPOSITE:

**Middle East: Oil Production.**

© Global Mapping

## 3-D symbols (volume)

3-D symbols work well when the data range is very large as the size of the symbol is proportional to the cube root of the data value. The symbols are drawn in 3-D so that they appear to have volume. This requires skill. In the case of proportional spheres, again estimate the size of the largest sphere that will fit on the map, then find the cube root of the highest data value and calculate the scaling factor as you did for 2-D symbols. You then need to construct a symbol that appears to have the right dimensions. Again, you may find that range grading the data is easier.

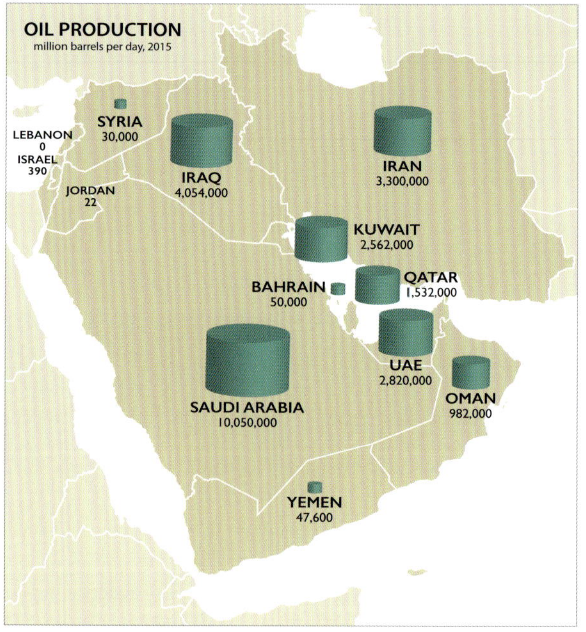


Legends for proportional point symbols should show a range of sample sizes, from low to high. Nest circles or squares to save space if necessary. Show the constituent parts of pie charts.

## Line symbols

Lines showing different quantities are varied in width or in colour; varying width usually works better. The basic principle is that the width of the line is varied according to the size of flow. Absolute or ratio data can be used, so they can show many different types of data. Direction is usually important and is indicated by using arrows. Flows that divide (literally or figuratively) should divide on the map.

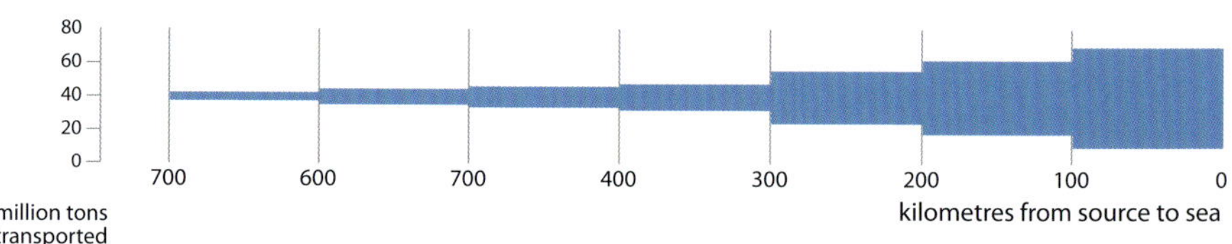

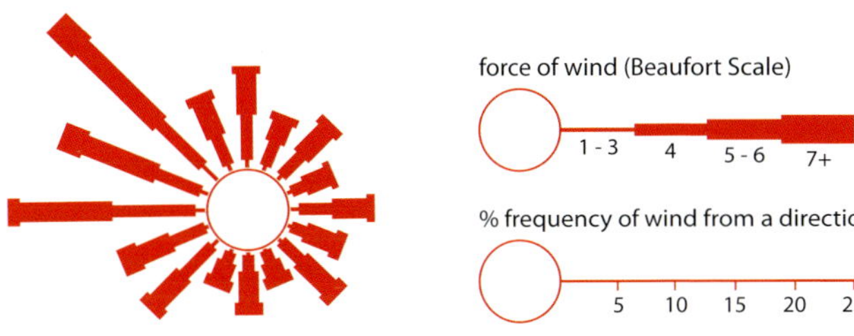



Constructing a quantitative flowline map involves working out the maximum width of line suitable for the scale of the map (often determined by trial and error). Wide lines look like areas, but that may be appropriate to give an impression of high activity. Find the highest value and lowest value items of data. The highest value will be represented by the widest line. Calculate the scaling factor, and apply it to all other data items, checking that the lowest value is still wide enough to be seen — probably no narrower than 0.25mm. Range grading data may be necessary to take into account very small or large data values.

Flow lines should be the figural element of the map, so they need to be on a higher visual plane than the base map. Adding data values to the lines can save the need for a comprehensive legend.

### useful tip

Where lines cross, put the narrower lines above wider ones, and use a hold-out to emphasise the narrower lines.

### useful tip

Adding a shadow to a flowline helps lift it above the base detail.

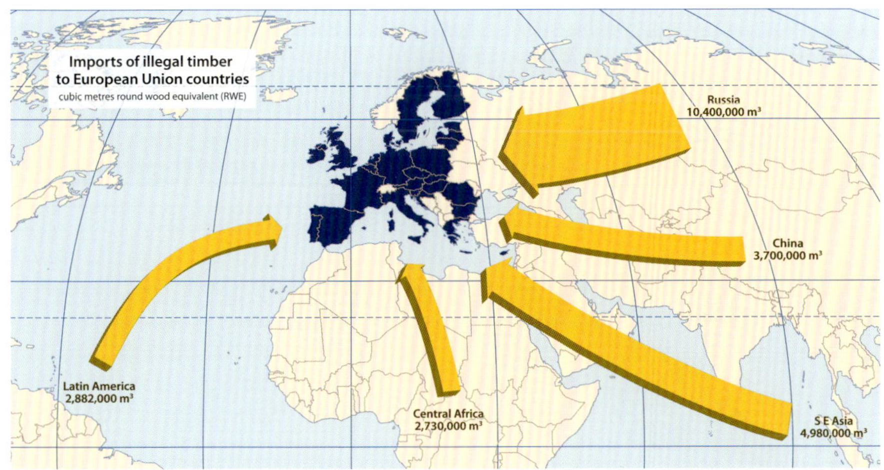



Environmental World:
Illegal Timber Imports.
© Global Mapping

## Areas

Quantitative area maps are usually choropleth maps or isoline maps with fills for different bands. The essential principle is that colours are varied either by saturation of one hue (colour) or variations of hues to reflect increasing data values. The subjects are explained in chapters 30 and 32.

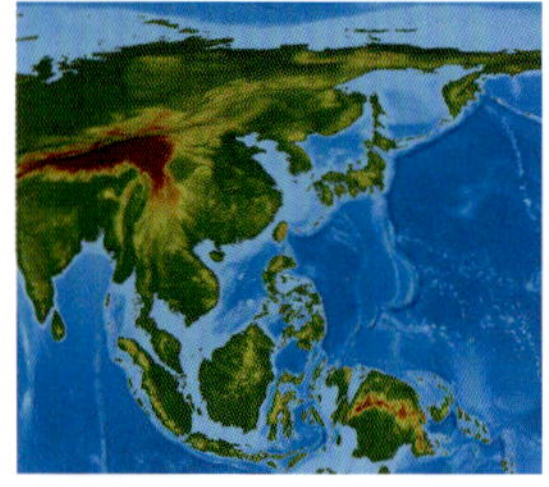

Gridded precipitation cartogram.

© Benjamin Hennig

Gridded population cartogram showing precipitation patterns: Each grid cell is resized according to the total number of people living in that area. The areas are then shaded by the annual precipitation in that area.

© Benjamin Hennig

Non-contiguous cartogram of the 2012 Presidential election: The size of the states is modified by number of votes for the winning candidate, which reduces the red and increases the blue.

© Kenneth Field

**N**OT ALL MAPS have to show real-world geography. Some of the most successful maps deliberately distort the real relationships between features in the interests of emphasising one important aspect of the theme being mapped. Transport diagrams are good examples of how real-world geography is deliberately distorted to aid movement.

Another example that has increased in popularity and is now often produced is the cartogram. Cartograms are also known as value-by-area maps because units from a base map are distorted so that each unit is shown in size in proportion to the value being mapped. For example, a cartogram of the world's population by country shows populous countries like India and China larger than their size on an equal-area map, and Australia shrunk, to reflect their populations relative to their areas. They are a form of combination of map and graph.

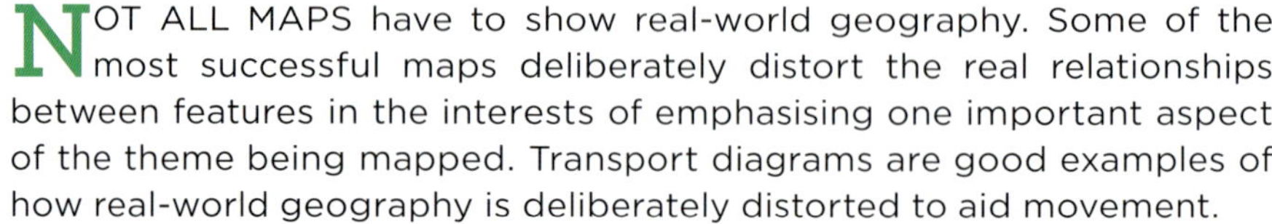

Cartograms can be effective ways of presenting data, revealing patterns not readily seen in more conventional mapping. They work best when a comparison of the data and the area of the unit being mapped reveals a reversal — a large data figure and a small unit, or vice versa.

There are several types of cartogram, but three types are worth exploring here. The non-contiguous cartogram takes each statistical unit (e.g. a country) and expands or reduces it in proportion to its data value. They are then depicted as a series of separate units on a map, placed in roughly the right relative positions but not touching. They have the advantage of maintaining correct shapes, but sacrifice the topological relationship between areas.

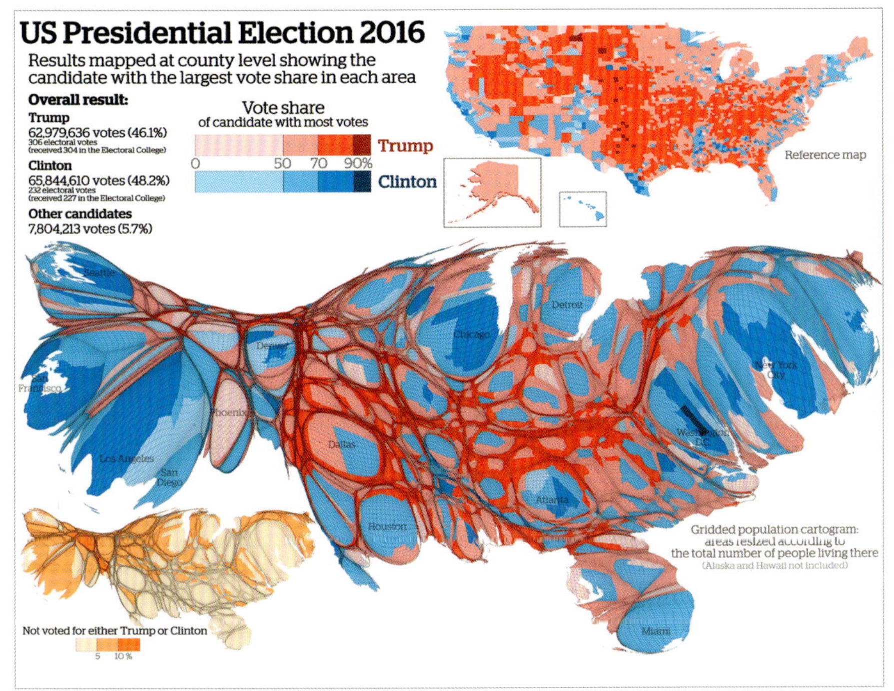



Gridded population cartogram of the US Presidential election 2016: Each grid cell is resized according to the total number of people living in that area. The areas are then shaded by the proportion of votes for each respective candidate in the election.

© Benjamin Hennig

Contiguous cartograms keep the topological relationships between units correct, but sacrifice the shape. For example, if two countries touch one another in reality, they must be shown as touching on the cartogram. Contiguous cartograms are now more common than non-contiguous because computer programming greatly helps in their preparation.

Dorling cartograms don't make any attempt to keep the shape of areas recognisable. Instead they use standard geometrical shapes (typically circles or hexagons) to represent the attribute or data value of each area which are then placed in a pattern that approximates to their real-world relative positions. It is in some ways like a proportional symbol map. The symbols do not overlap one another.

Dorling cartogram: Animation of obesity in the USA, 1995 to 2008.

Courtesy: Stanford Visualization Group

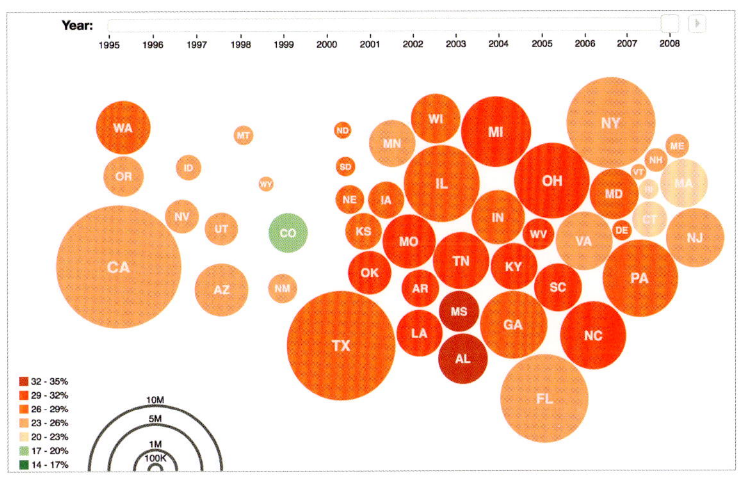

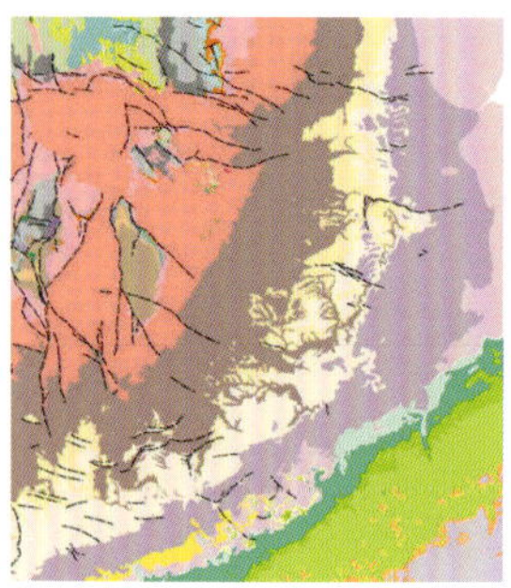

**Digital Geology 1:625,000: Created from online open source data.**



*A succint title, subtitle and note (far right) is so much better than the wordy description.*

MANY OF THE MAPS produced from a GIS query using the default settings don't convey a message well because the cartography is poor. Default maps often take little account of the essential cartographic principles of legibility, good structure, correct use of colour and pattern, visual contrast, the figure-ground relationship, balance and good typography. Many problems come about by the use of too much data on the map, too many classes, and the poor use of colour.

Most GIS maps can be improved with a little work. If you have sophisticated cartographic tools as part of the GIS program, then use them. There are also many open-source GIS applications available that offer great functionality and design capabilities. A map file can be exported to a graphics program like Adobe® Illustrator® where you have much more design flexibility. Programs like MAPublisher can act as a bridge between GIS and graphic design.

Along with other chapters, the checklist here can be used to suggest aspects of a GIS map that might be improved with a little thought.

## A checklist for better mapping

### Titles

Devise your own, meaningful title and omit the word 'map'. Usually the file name is meaningless as a map title.

Map of the British Isles showing number of people living in each square km for 2016

**BRITISH ISLES**
Population Density
people per sq km, 2016

### Colours

Default colour ranges are often poor, particularly for choropleth maps. Good maps often use subtle, balanced colours, and bold colours can look inappropriate. There are many suggested colour schemes for choropleth and qualitative maps at www.colorbrewer2.org.

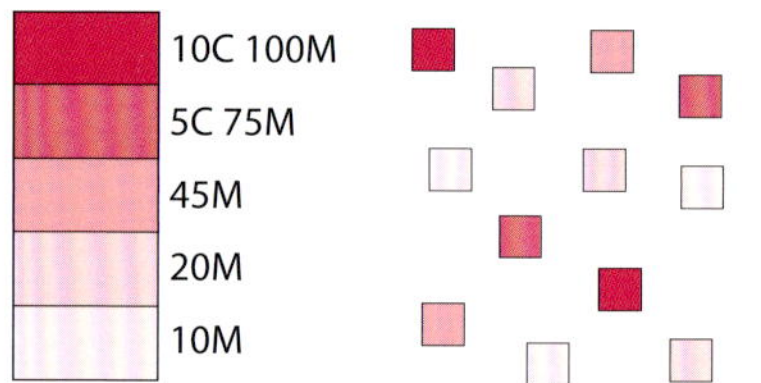

*A single colour progression may appear clear in a legend but when the colours are scattered randomly they become difficult to differentiate.*

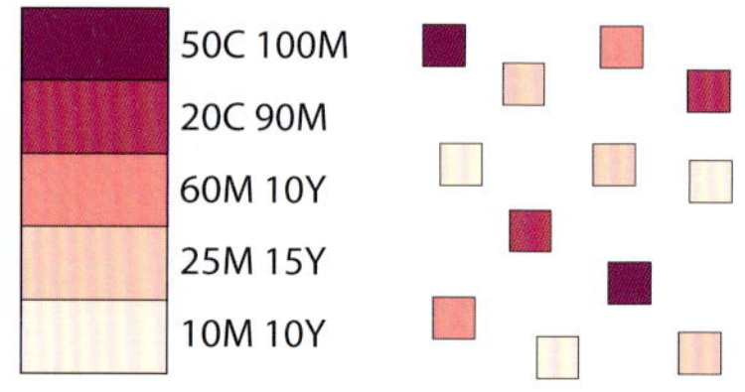

*By increasing the cyan at the top of the range and adding yellow at the bottom the colours are more easily identified.*

## Data classes

GIS queries often result in a huge number of data classes, but map users can only distinguish 6 or 7 colours and only 5 tones of grey with ease. Reduce the number of classes to between 4 and 8. Most data will need classification, but equal intervals rarely reveal much — get a feel for the statistics. Consider different classification options such as arithmetic or geometric series, using natural break points, and use logical break points (like zero). Ensure that the numbers used as break points between classes are meaningful; a fraction of a person is usually a challenging concept, and odd decimals are not intuitively understood.

> **useful tip**
> State the method of classification used (e.g. 'quantiles'), so that map readers can understand how the data have been classified.

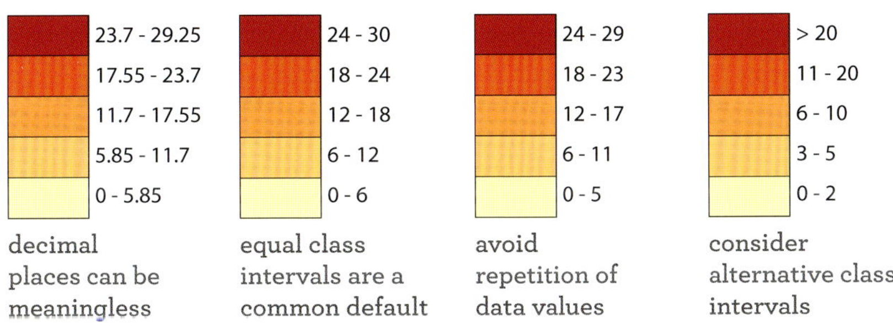

| decimal places can be meaningless | equal class intervals are a common default | avoid repetition of data values | consider alternative class intervals |

## Type

Poor type placement, inappropriate type sizes and poor default fonts make a map unreadable. Move labels that clash so that they can be seen, and edit out unnecessary labels. Many poor maps use inappropriately large map labels. Vary the map labels to show hierarchy of importance.

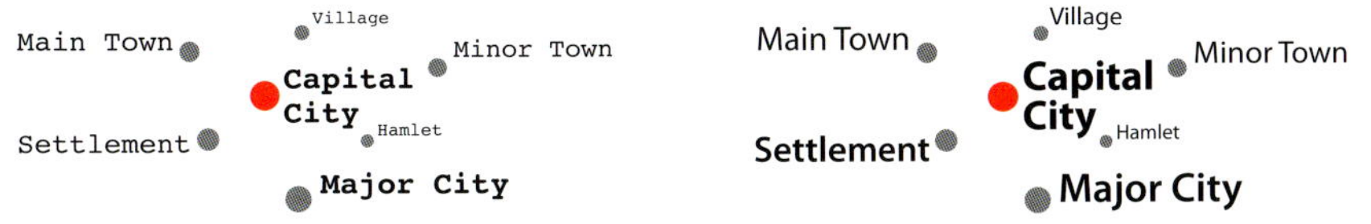


Names in Courier (left) need to be smaller to occupy the same space as in Myriad and are consequently much harder to read.

## Symbols

Check the colour and size of symbols. All symbols should be clear, legible and distinguishable from one another.

Smaller symbols require stronger colours and perhaps keylines as well to tell them apart.

## Legend

A legend should explain all symbols shown on the map. Group together similar items and similar symbol types (lines, point symbols, etc). Check that data labelling is clear and correct, and that the units and dates are shown (e.g. Density per km$^2$, 2017).

EVEN IF YOU HAVE CHECKED and checked again it is always worth one final check before the map is signed off. The more you work on a map the less you will see the errors — it is the obvious ones that will get through, such as the misspelling of familiar words. Here are some tips for improving your maps in 5, 15 and 50 minutes.

> **useful tip**
> Pay attention to detail and allow yourself enough time to get things right.

### In 5 minutes...
Check...
- the legend — do all the symbols in your legend match the map?
- the scale of the map — is it correct?
- your map title — is it brief and directly related to the map?
- the insets — does each have a title and scale bar?

### In 15 minutes...
As well as the above, check...
- the aspect ratio — will the map fit the screen with minimum scrolling?
- the legend units — does your legend state the right units (e.g. density per km$^2$) and correct dates?
- the spelling — check the spelling of place names
- labels — are they all distinct and separate?

Colour Vision Deficiency (CVD) affects many people — consider what adjustments can be made to your colours to ensure that the information being presented is meaningful to the widest possible audience.

### In 50 minutes...
As well as the above...
- print it out and proof it — mistakes are easier to spot on a printout
- if it's a web map, try it on different browsers
- get a second opinion on your map design
- ask someone else to check the spelling
- take one final look at the layout to make sure everything is in balance.

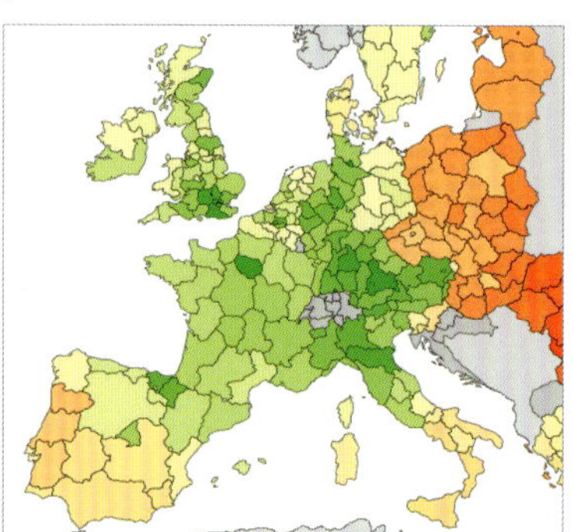

Original red-green colourway

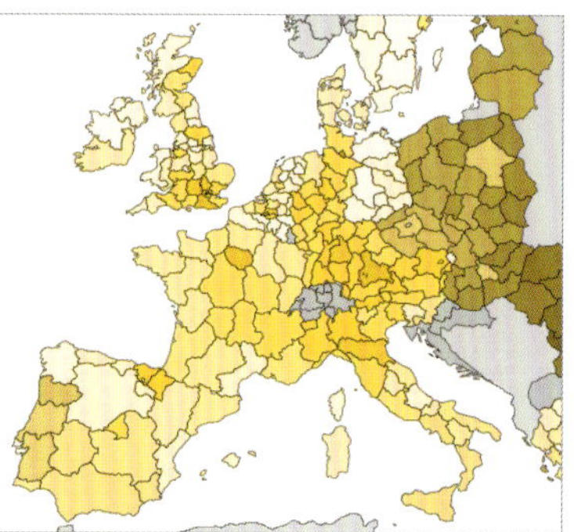

Simulation of protanopia, also known as red 'blindness'

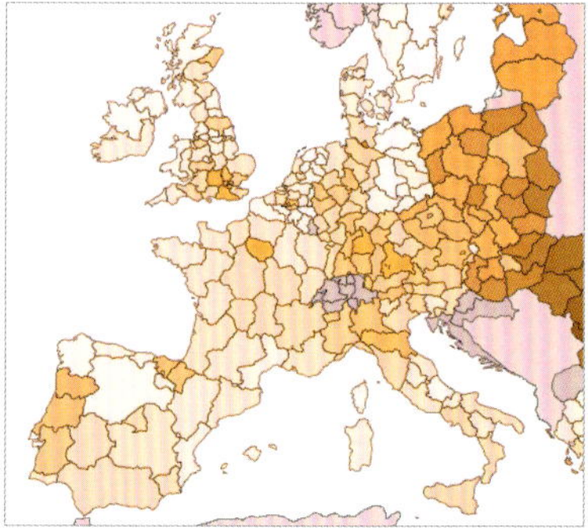

Simulation of deuteranopia, also known as green 'blindness'

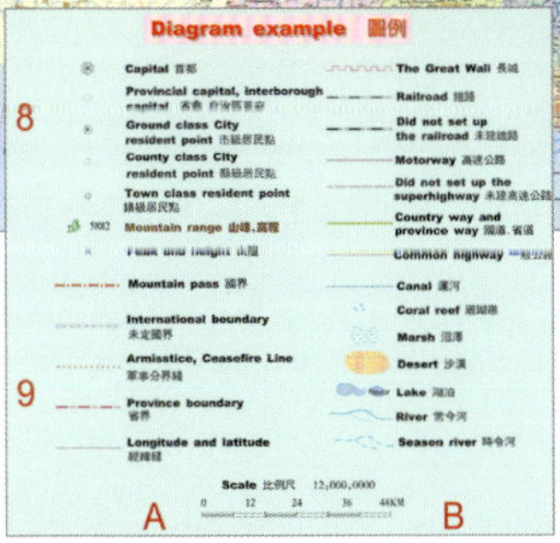

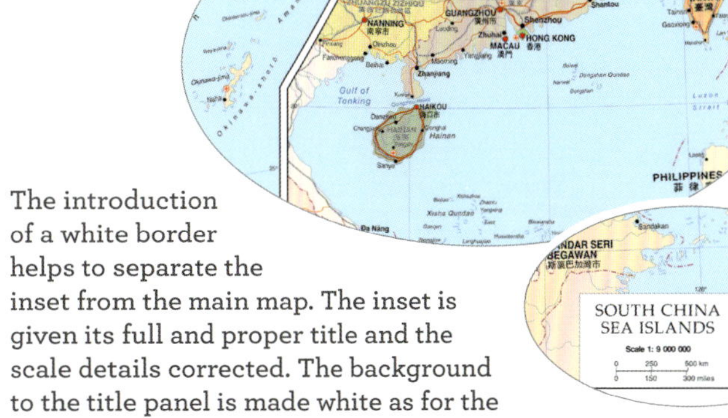

A few simple changes
can improve a poor map.

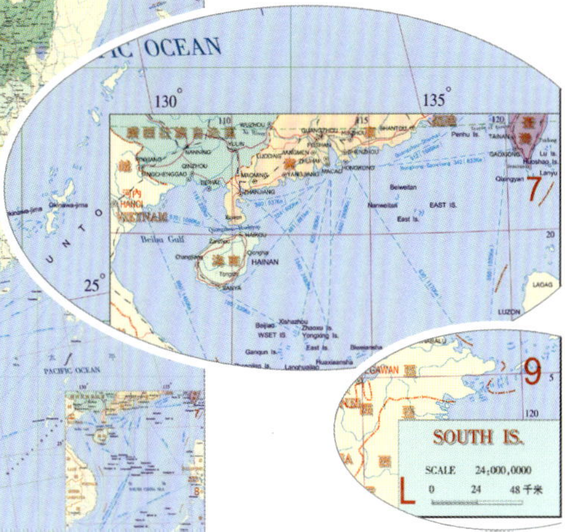

The original map (above) features a rectilinear grid as well as lines of latitude and longitude. The grid has been removed together with the grid references since there is no need for either in the absence of an index. When a grid reference is required the figures and letters should be placed outside the legend.

The very poor layout of the legend (above) has been improved and the superfluous heading removed. The title of the map is shortened and moved to sit above the new legend with the scale bar (below).
Note the scale ratio on the original map is given as '1:12;000,0000' — it should have said 1:4,500,000.

The introduction of a white border helps to separate the inset from the main map. The inset is given its full and proper title and the scale details corrected. The background to the title panel is made white as for the main legend for easier identification.

# Acknowledgements


Agency for Data Supply and Efficiency, . . . . . . . . . . . . . . . . . . . . . . . . . . . . 19
www.sdfe.dk

Applied Wayfinding . . . . . . . . . . . . . . . . . . . . . . . . . . . . . . . . . . . . . . . . 13, 14
www.appliedwayfinding.com

Benjamin Hennig. . . . . . . . . . . . . . . . . . . . . . . . . . . . . . . . . . . . . . . . . 88, 89
www.viewsoftheworld.net

British Antarctic Survey. . . . . . . . . . . . . . . . . . . . . . . . . . . . . . . . . . . . . . 6, 71
www.bas.ac.uk

British Geological Survey . . . . . . . . . . . . . . . . . . . . . . . . . . . . . . . . . .64, 70, 90
www.bgs.ac.uk

Clear Mapping Company. . . . . . . . . . . . . . . . . . . . . . . . . . . . . . . . . . . . . . . .6
www.clearmapping.co.uk

CollinsBartholomew . . . . . . . . . . . . . . . . . . . . . . . . . . . . . . . . . . . . . .8, 57, 77
www.collinsbartholomew.com

Communicarta. . . . . . . . . . . . . . . . . . . . . . . . . . . . . . . . . . . . . . . . . . .15, 36
www.communicarta.com

Cosmsographics . . . . . . . . . . . . . . . . . . . . . . . . . . . . . . . . . . . . . . . .1, 6, 70
www.cosmographics.co.uk

Danish Geodata Agency . . . . . . . . . . . . . . . . . . . . . . . . . . . . . . . . . . . . . . 19
www.eng.gst.dk

DataShine. . . . . . . . . . . . . . . . . . . . . . . . . . . . . . . . . . . . . . . . . . . . . . . 1
www.datashine.org.uk

Europa Technologies . . . . . . . . . . . . . . . . . . . . . . . . . . . . . . . . . . .12, 25, 69
www.europa.uk.com

Garsdale Design . . . . . . . . . . . . . . . . . . . . . . . . . . . . . . . . . . . . . . .6, 12, 66
www.garsdaledesign.co.uk

Geographx NZ. . . . . . . . . . . . . . . . . . . . . . . . . . . . . . . . . . . . . . . . . . . . .8
www.geographx.co.nz

Global Mapping. . . . . . . . . . . . . . . . . . . . . . . . 6, 48, 52, 57, 76, 81, 82, 84, 85
www.globalmapping.uk.com

Harvey Maps . . . . . . . . . . . . . . . . . . . . . . . . . . . . . . . . . . . .1, 9, 18, 28, 47
www.harveymaps.co.uk

Insurance Services Offices, Inc. . . . . . . . . . . . . . . . . . . . . . . . . . . . 19, 25, 28, 47
www.verisk.com/spatialdata

Imray Laurie Norie and Wilson. . . . . . . . . . . . . . . . . . . . . . . . . . . . . . . . . . 40
www.imray.com

Numerous examples of mapping that appear throughout this book have been specially adapted or created by Mary Spence MBE from data supplied by XYZ Maps and maps published by Global Mapping. Special thanks must go to both companies for permitting their work to be used in this way.

The sketches employed to illustrate cartographic principles have been prepared and supplied by Mary Spence MBE.

# Further information

Some useful further reading:

**Designing Better Maps** A Guide for GIS Users by Cynthia A. Brewer
*ESRI Press, Redlands, California*

**Making Maps** A Visual Guide to Map Design for GIS by John Krygier and Denis Wood
*The Guilford Press, New York & London*

**The Routledge Handbook of Mapping and Cartography** edited by Alexander J. Kent and Peter Vujakovic
*Routledge, Abingdon*

**The Cartographic Journal** is the British Cartographic Society's peer-reviewed, international periodical and publishes the latest research in the world of mapping.

There are many online resources to help with the selection of colours, some of which are specifically created for mapping use:

- ColorBrewer 2.0 at colorbrewer2.org – especially useful for thematic mapping offering colour schemes for different classes of data
- Adobe Color at color.adobe.com – a colour scheme generator which includes analogous, complementary and other schemes plus a vast array of existing themes to inspire
- Paletton at paletton.com – a tool for creating colour combinations that work well together including simulations for colour vision deficiency.

In addition, there are a number of online tools that test the viability of colour choice by simulating colour-impaired vision, such as:

- Color Oracle – colororacle.org
- Coblis – color-blindness.com/coblis-color-blindness-simulator/
- Vischeck – vischeck.com
- Adobe Photoshop has a proof setup option to view the image in protanopia- and deuteranopia-type simulations.

The British Cartographic Society is a registered charity open to all with an interest in maps and cartography.  Its aim is to promote all aspects of cartography in the UK and abroad by offering a forum for the exchange of ideas and the sharing of cartographic knowledge.

Details of activities and membership can be found on the Society's website **www.cartography.org.uk**.